# Proteins Crossing Membranes

# Proteins Crossing Membranes
## A Scientist's Memoir

Stephen Rothman
University of California, San Francisco

**CRC Press**
Taylor & Francis Group
Boca Raton London New York

CRC Press is an imprint of the
Taylor & Francis Group, an **informa** business

CRC Press
Taylor & Francis Group
6000 Broken Sound Parkway NW, Suite 300
Boca Raton, FL 33487-2742

© 2019 by Taylor & Francis Group, LLC

CRC Press is an imprint of Taylor & Francis Group, an Informa business
No claim to original U.S. Government works

Printed on acid-free paper

International Standard Book Number-13: 978-0-367-07450-0 (Hardback)
International Standard Book Number-13: 978-0-367-07449-4 (Paperback)

### Library of Congress Cataloging-in-Publication Data

Names: Rothman, S. S. (Stephen S.), author.
Title: Proteins crossing membranes : a scientist's memoir / Stephen Rothman.
Description: Boca Raton : Taylor & Francis, 2019. | Includes bibliographical references.
Identifiers: LCCN 2018040072 | ISBN 9780367074494 (paperback : alk. paper) | ISBN 9780367074500 (hardback : alk. paper) | ISBN 9780429020810 (General)
Subjects: | MESH: Membrane Proteins—metabolism | Biomedical Research | Cell Membrane Permeability | Transport Vesicles | Peer Review, Research—ethics | Personal Narratives
Classification: LCC QP552.M44 | NLM QU 55.7 | DDC 572/.696—dc23
LC record available at https://lccn.loc.gov/2018040072

**Visit the Taylor & Francis Web site at
http://www.taylorandfrancis.com**

**and the CRC Press Web site at
http://www.crcpress.com**

*To students, fellows, and colleagues who have been companions on this perilous journey and to all those scientists, past, present, and future who find themselves at odds with the common view.*

# Contents

*Nullius in verba*

# Foreword

(Nullius in verba *is the motto of the Royal Society, London. Its members chose it in 1660 soon after the Society's formation. It is Latin for "on the word of no one" or "Take nobody's word for it," and is taken from Horace's Epistle. The Society explains it as "an expression of the determination of Fellows to withstand the domination of authority and to verify all statements by an appeal to facts determined by experiment."* Proteins Crossing Membranes *tells the story of my commitment and that of my students, fellows, and colleagues to this injunction as we found ourselves in a clash with a central belief of cell biology.*)

In an intriguing series of experiments carried out many years ago, a common scientific belief, feted by no less than three Nobel prizes, was brought into question. The observations were about proteins—the molecules that the genetic code specifies and that are in one way or another central to all of life's activities. The experiments however were not about what proteins do, but how they are moved, in particular how they are moved from where they are made to where they act.

The results of these studies were in conflict with the standard view of how this happens, and thus became controversial. Though controversies happen all the time in science, there was an enormous one might say an extraordinary problem in this case. Scientific controversies are usually resolved by test. How well do the different points of view acquit themselves when faced with the properties of the natural world?

This was not the case here. Those who believed in the standard theory were *not* interested in testing it against the properties of nature. Their interest, seen in all their copious research, was to do one thing and one thing only, to bolster their theory's dominion. Negative observations were ignored or dismissed. They did not as one might expect lead to greater scrutiny of the standard view. To the contrary, time and again, obdurate belief trumped observation, leaving the entrenched perspective even more entrenched.

Supporting this view were many distortions and manipulations of data, including the conflation of interpretation and observation, with hypotheses being presented as facts of nature. In telling this story, *Proteins*

*Crossing Membranes* provides a cautionary tale. Though the teaching and practice of science is often celebrated as an oasis of reason and rational thought amidst the deep shadows of intellectual and moral confusion of the humanities, skepticism is prudent even when the claim is scientific.

**Stephen Rothman**
*Berkeley, California*

# Preface

Richer than ever, sought out by students to gain a foothold on the future, the American university is in a sad state. The traditional liberal values of open inquiry, diverse opinion, and welcomed debate is under attack by a pernicious political correctness that not only expunges free speech, but punishes those harboring views deemed offensive, improper, and even indelicate by aggrieved students, all this with the complicity if not the outright support of the professoriate and feckless administrators and boards. As Allan Bloom told us over a quarter of a century ago in his *The Closing of the American Mind*," higher education has failed democracy and impoverished the souls of today's students."

None of this has been accidental or due to inattention. Powerful weapons have fueled the challenge. The weapons, Bloom pointed out, are critical and moral relativism. Who is to say what something means or what is moral or immoral? It is all relative, just a matter of opinion. Who are you with your deeply flawed self and culture to judge others and their culture, no less impose your sense of meaning and morality on them?

This may simply seem like a call for modesty, except that the authors of political correctness are far from modest. They may say that it is all a matter of opinion or perspective, of judge not, lest thee be judged, but they sanctify their own analysis and beliefs as being unmistakably true and moral, and condemn disagreement as either false, immoral, or both. Much of this seems to have been driven by a desire to tear down the old order, whether in the university or society, and replace it with a new "fairer" (even, utopian), more inclusive and diverse system of the critics' imagination.

Whatever you think of this critique of the modern American university, wherever your sympathies lie, this appraisal, whether attacking or defending, has been almost exclusively focused on the "humanities." Science, its teaching and practice, is almost invariably excluded from the clash. Indeed, it is, as said, often celebrated as an oasis of reason and rational thought, of open and dispassionate analysis, a bright light and clarion call amidst the deep shadows of intellectual and moral confusion of the humanities.

But is such praise justified? Is science free of moral and intellectual confusion? Putting aside politically tinged science, such as research on the environment, is modern science, even pure science, science for its own sake, a benevolent home for open and honest debate, a place for deep analysis and contemplation where differences of opinion are welcomed, even celebrated, or is this wishful thinking or worse, a conscious deception?

This is the subject of this book. I approach the question through my own experience. As such, it is a memoir, though it is an unusual memoir in that beyond some brief personal portraits, the chronicle is mostly about science, not people. Even though it touches on what I view as my successes, disappointments, and apprehensions, and what I perceive to be the good or bad conduct or inclinations of others, as well as more generally on the character and disposition of the purveyors of science, the investigators, and the system that supports their activities, at its core the story is neither personal nor political. It is about science, about laboratory research, and its interpretation. In particular, it focuses on fundamental research on the biology of the cell as it has been carried out in modern times.

Critical to the discussion is the deference or lack thereof that scientists in this field have given to the actual properties of nature. Theories and models have been constructed based on what scientists imagined is true or wished was true, not on what nature decreed or ordained. It was in this failure, I would say this abject failure that my youthful romance with science confronted reality.

The story, told from the perspective of my old age, is simultaneously discouraging and filled with beauty and hope. It starts when as a young scientist I found myself in an intense and disturbing controversy, we can call it a disjunction, between observations made in my laboratory and an entrenched and widely held point of view that was at the time and remains central to modern ideas about the character and properties of the biological cell. It seemed that what others believed, honored by three separate Nobel prizes by now, and what I observed could not both be true. Either their beliefs or my observations were wrong. In *Proteins Crossing Membranes*, I chronicle the controversy and the science behind it.

At first, others viewed my observations with curiosity and interest. But before long, the research came under fire, especially from those with an investment in what was already the standard point of view. My results just could not be believed. When they turned out to be accurate, a reflection of nature's properties, they were either ignored, noted with biting contempt, or it was pointed out that not only could they be accounted for by the standard model but also that without doubt this was the case. Behind the scenes, it was whispered that the work was poorly done, poorly thought out, or even fabricated. I became a heretic with suspect, if not base motives, a bad scientist, at best incompetent, at worst up to no good.

In all this, the critics thought themselves fair and open-minded, judging the facts as they were, without bias or prejudice. In high dudgeon they would say, with consummate scientific modesty, "Certainly our model might be incorrect, all scientific concepts can be wrong. But at the same time, it is substantiated by powerful evidence and we are confident that it provides a faithful rendering of how things actually happen."

As the negative evidence grew, minds were not opened, but closed. Adherence to the standard model became more, not less rigid and unyielding. The wagons were circled, and the battlements reinforced to defend against encroaching doubt. What good could come from tarnishing brilliant experiments by brilliant scientists? What was needed was to build on what had already been accomplished, not to destroy or disparage it. And so it was that a plausible model became dogma, truth incarnate.

I was an outsider. My prior research had nothing to do with the model or even this field of study, and in the beginning, I did not doubt the common wisdom of the experts. Indeed, I expected my experiments and those of my students to bolster their claim, and dreamt of getting at least a modicum of credit for our contribution. But things did not work out this way or so genially. Our results fed doubt; they did not assuage it. Nature, it seemed, had its own ideas, and it held all the cards. The answers it provided just made things worse. The unwanted controversy would just not go away. Nothing we found strengthened the standard view.

And so I became a critic of the model, at first uncertain and hesitant, but before long full-throated, yet always hoping against hope that I was wrong and that I would be able to convince myself that the standard view was correct. Though of a skeptical and intellectually aggressive nature, I was also very naïve and completely unprepared for the storm on the horizon. I believed that, as scientists, others would with time accept what I had exposed and that, as truth seekers, they would modify their view and if necessary reject their own model if nature demanded it. However discomforting and whatever their predisposition, they would, I was sure, submit to nature, to *its* properties. At the time I was not in touch with, just how very unlikely this was, with how abounding the prejudice that colored everything was.

I expect that, even after all these years on learning of this book, advocates of the standard view will either ignore it or condemn it with scathing derision. They might with ersatz gentility dismiss it as "old hat" and say that the time has long since passed for me to accept the simple fact that I was wrong. Less genially, they might add, stop beating a dead horse! Your observations were interesting, even provocative at first, but they have long since been shown to be false or accounted for by the standard model.

To continue to raise this ancient ghost is no more than obsessive blather. The facts *have* been established and the standard model validated as any reasonable assessment of the evidence affirms. It is no longer a just model, but a certified fact of nature. It could not be more secure.

But like my younger self, even now, in my dotage, I demur. I disagree with my imagined critic. The properties of nature that conflict with the model, have not come anywhere near being paid their full due. Fealty to nature's truth, acceptance of its dominion however unwelcome is not a choice for a scientist but an imperative. There are no exemptions for fine ideas and sophisticated evidence, no less for widely held beliefs. The simple fact is that the observations we made all those years ago, as well as others that have surfaced since, have never truly been accounted for. They were just ignored or explained away—sidestepped, eluded, avoided, and when noticed, alternative explanations consistent with the standard view not merely imagined, but as said, assumed to put an end to the issue. But the fact of the matter is that these experiments and the data they produced remain unsightly, unerased, and I believe unerasable blemishes that tarnish the pristine countenance of the standard model's account of nature.

Of course, scientists, like all of us, can ignore or dismiss what they find jarring and uncomfortable if that is their disposition. But as long as relevant facts remain unexplained, discounted, or disregarded, the claim of disambiguation rings hollow. In this, it is not acceptable to merely claim, however strenuously, that a model is true, that it provides an accurate account of nature's properties without exposing it to the possibility of failure, to experimental test.

This is the crux of the matter. Our observations were not merely what we happened to see in the laboratory. The properties of nature we uncovered were *sought out* to test the standard view, to test the acclaimed theory against nature's attributes. How well did it acquit itself when faced with properties of the process it was said to explain? Was it explanatory or was it not?

Against our observations, disciples of the standard model will doubtless point to the mountain of evidence that supports their viewpoint as proof of its correctness. But the fact is that mere support or corroboration for a particular point of view, however voluminous, however detailed and long standing, is insufficient if the truth is what we seek. It does not matter how many affirming studies there are, how explanatory they seem, or how impressive the results, if the experiments do not *test* the theory's validity against the properties of nature, the assertion remains unproven. A true portrayal of nature can only be claimed for a model if it has survived the fire of experimental and theoretical test, tests that it might have, but did not fail. Our efforts aside, such tests were simply not performed.

This was very discouraging. It was a dreadfully dispiriting statement about science. Large swathes of research sought to establish the hegemony of the standard model, but never tested it either in its own right or in comparison to alternative explanations. Indeed, as said, evidence that conflicted with the entrenched belief was invariably dismissed, discounted, or ignored. Though vehemently denied, biases were and remain plentiful to this day.

Nor were the stains on the established understanding so trivial that they could be deemed insignificant. The discordant observations were not only inharmonious with the explanation provided by the standard model, but also opened the door to a wholly different account of the experimental data. This is the aforementioned disjunction. Our observations not only offered another way of looking at the process, but also gave credence to what turned out to be the *sole* alternative to the standard view. Moreover, that alternative was its antithesis. It did not merely differ; it was contradictory. Even more disquieting, based on first principles, it was more desirable. Though under certain circumstances, the two points of view could exist side-by-side, as you shall see, for clarity, we sought evidence that could distinguish one from the other, which could show that things either occurred this way or that, not both ways.

In the end, the simple fact is that unless we test our ideas against the properties of nature, whatever we believe to be true and however passionately, all we are doing, all we can do is expound and elaborate a dogmatic creed. Testing our theories against nature is crucial if we wish to claim that they are scientific theories. Putting all this aside, the story I am about to tell is not only uplifting, but it is also one of undeniable beauty. It tells of one of humankind's greatest gifts. For reasons beyond anyone's comprehension, however deeply flawed, however morally and intellectually deficient we humans are, only our species is capable of exposing the properties of nature. All other living things merely live by and act in accordance with them. We alone are capable of thinking about and at least attempting to understand their underlying nature.

To me, this gift of the intellect is not only about intelligence but also about beauty. Indeed, in a sense it is the deepest beauty of all, the beauty of discovery. When, as a young unfocused student, I was first exposed to nature's truth, to its properties by experimentation, my way of looking at life and nature and my personal intersection with the physical world changed forever. I was overwhelmed, humbled, and awed by the splendor and the exquisite beauty of uncovering its mysteries.

Why advocates of the paradigm have not properly, not to say aggressively, exposed their model to tests that it might fail in the some 60 years since its original proposal is not for me to say. For those steeped in its beliefs, certainly nothing written here is likely to change their frozen minds. Yet, I insist today as I did all those years ago that the incompatible observations remain facts of nature that must be fully addressed, not explained away.

Whether science is self-rectifying, whether the truth will sooner or later be paid its full due is an open question for me. I wish I could say that I have confidence that a reckoning will come and that the incongruous observations will eventually be taken into account, but I cannot. They may simply disappear into the ether as if they never existed, and the dogma will remain in place and even become strengthened. I believe that prospects for the truth are not a matter of the abstract concept we call science. It is inherently and manifestly about human nature. Finally, given my age and whatever the ultimate fate of the observations I will discuss here, this is probably my last chance to tell you about them.

A great challenge in telling this story has been the awkward tension between my desire to write a book that is accessible to the nonexpert and that as such would not be technical, and being true to the authentic character of the research. I could not provide all the detail needed to make the various arguments as strong as possible, because if I did, what I wrote would be obtuse to all but the expert. On the other hand, if I provided too little information, the case would not be made, and the basis for the conclusions I draw would be unclear.

At each and every turn, I had to ask, am I saying too much and thereby confusing things, or am I confusing things by saying too little? Is the example I have chosen apt and is it comprehensible? Add that the subject matter is complex and at times a little difficult to understand no matter how I put things, and perhaps you can appreciate my struggle. I hope that I have been at least partially successful in walking this tricky line. Where I have failed, you have my apologies. *Proteins Crossing Membranes* faces another challenge. I just read a review of a book on the history of science's understanding of reproduction. The reviewer congratulates the writer for highlighting the many discoveries and brilliant insights of scientists and thinkers over the millennia from the Greeks to 19th century Europe that led to our modern understanding. But he ends his review by especially thanking the author for pointing out the role of "unexamined assumptions, ad hominem arguments, misplaced ego, and stubborn adherence to outworn theories" in hindering progress.

When historians tell us about the foibles of the checkered past of scientific discovery, we, besotted by modernity, by modern science, smile contentedly. We are free from such impediments. There are no "unexamined assumptions, ad hominem arguments, misplaced ego, and stubborn adherence to outworn theories" to hinder progress today. These are vexations of days long past. Modern science's understanding of nature is substantial, safe, and sound. What we know, we know securely.

But *Proteins Crossing Membranes* has the temerity to tell us that such impediments have not vanished. That obscurantism, in all its various forms, is alive and well in the science of our time. This is a formidable task for a memoir. Memoirs are about relatively recent events, not ancient history, and it is far easier to criticize events and attitudes of the remote past, than those of our own time. Even though facts may be few and fragmentary and large dollops of inference may be required to imagine what actually occurred, historians have the license of distance. On the other hand, the memoir, even though providing an actual account of events, necessarily does so through the eyes of a participant and through the lens of personal attitudes. Though the distortions can be moderated or minimized, they can never be completely eliminated. But these two tellings of a story, one of history and the other of experience are not really in conflict, but are complementary. As historians seek to piece events from the long distant past together, they often rely on first-person, firsthand accounts from the time to help them construct their narrative.

Except to note the contributions of some of my students, I have used the names of the people involved, heroes and villains alike, sparingly in an attempt to inoculate the story from being seen as an expression of personal pique, of getting even with adversaries and critics. At base, the tale is not about them or me. It is not about personal feelings and attitudes, though of course, they were and are plentiful. As such, it is not about bile and its color. It is about science, what is foul and ugly about it, as well as what is ennobling and beautiful.

# Acknowledgments

Whether you welcomed the work or found it frightening, I want to express my deepest gratitude to all of you who in one way or another participated in this struggle, to all my students, fellows, and colleagues. I would also like to thank those few outside our small world who did not turn a cold shoulder, but were welcoming and supportive in what were often difficult circumstances. A person whose name does not appear on the book's pages, Helga Wilking, has my abiding appreciation and praise for her crucial role, commitment, and above all her care and fidelity in the work. Finally, I would like to give my special thanks to Chuck Crumly for his interest in the project, to Jennifer Blaise, as well as to Judith Simon and all the others at Taylor and Francis for their help, and to Esther Saks for her skillful editing.

# Praeludium

From the Preface, Rothman, *Lessons from the Living Cell* (McGraw/Hill, 2002):

"Those who equate science with progress, and those who see it as a great danger to the earth and to themselves, often share a common view of what science is and how it works. In the broadest sense, and whether for good or evil, science is seen as a rational, unbiased or objective means of gathering and interpreting facts about nature in order to provide a clear and rigorous understanding, and eventually mastery, of its properties.

As any scientist knows and as beginning graduate students find out quickly, this is an idealization. When novice graduate students first learn to carry out and interpret their own experiments, they quickly come face to face with the real world of science. Facts, they find, are often malleable and uncertain, not incontrovertible, and interpreting them is quite often less a matter of applying meticulous reason than providing an opinion. And when they look beyond their own research to the broader evidence in their particular field, they find the same uncertainty and ambiguity in the work of others....

Experiments, observations, and their interpretation do not exist in some splendid, hermetically sealed world of reason. They are embedded in and are the consequence of a culture that, although scientific, is replete with biases, prejudices, and, significantly, all sorts of suppositions. Scientists may choose to believe certain results but not others, or to accept a particular interpretation but not another, and not solely because reason demands it of them.

As a community, scientists come to a shared sense of what is known and understood, what remains to be learned, how to learn it, and, finally, what is possible and what is not. ... Thomas Kuhn, in his classic book, *The Structure of Scientific Revolutions* (University of Chicago Press, 1962), called such shared beliefs paradigms, a word he borrowed from the social sciences and that has long since entered our everyday vocabulary. At any given time, he said, experts in a field of science tend to hold a common and comprehensive view of the system under their scrutiny—the paradigm.

Kuhn painted a portrait of science as a series of such paradigms (held by different groups of scientists), supported by evidence and for good reason, but also in varying measure by biases, assumptions, suppositions, rationalizations, opinions, prejudices, fallacious reasoning, personal animosities, political considerations, and every other human trait imaginable.

The paradigm represents a shared value judgment about nature, and value judgments are rarely if ever the product of evidence and reason alone. Kuhn realized that new evidence, however compelling, did not necessarily move scientists, like so many yielding reeds in nature's wind, to alter long-held and respected views. Quite the contrary, contradictory evidence often left older views more deeply entrenched, less open to question.

... For some students, coming to appreciate these difficult facts of scientific life can be demoralizing, and many a promising scientific career has ended before it began because of them. How can I proceed in the face of such uncertainty and ambiguity? How can I ever learn the truth if all I can do is seal the holes in my knowledge with shared presumptions about nature?

Kuhn said that most scientists work within a particular paradigm. He called this normal science. The task of normal science is to seek evidence to bolster the paradigmatic view, and to interpret whatever is found in its terms. The bulk of scientific progress occurs in this fashion, as new things are learned and fit to prior belief, proceeding by means of small questions and modest steps. Kuhn argued that major shifts in our understanding of nature, on the other hand, come about only when the foundational tenets of a discipline are seriously challenged: when evidence is sought not to bolster the paradigm, but to dispute it. Such new understanding, he argued, comes about by sea change, by cataclysm, indeed by revolution, not by some smooth and reasoned process of gradualism.

But who would produce the revolution? Who would break the covenant of understanding among a community of knowledgeable experts? After all, that covenant reflects their common perception and hard-earned wisdom. How could questioning at a fundamental level occur in such a situation? Kuhn points to outsiders, individuals who, ignorant of the paradigm and its understandings, unwittingly bring its foundational beliefs into question. Whatever the cause, when this happens—as it gratefully does from time to time—the established order is threatened, and that threat may be seen as no less than a challenge to reason itself. Confronting the menace, supporters of the paradigm defend their position strenuously through research and argument. They attack the contradictory evidence, its interpretation, and sometimes the individuals who present it. And as they do, Kuhn's revolutionaries continue to hurl rocks at the edifice as it is in the process being fortified, to see how sturdy it really is.

This is science as a kind of war; a war of evidence and ideas, with its tools of logic and reason all right, but also with weapons of passion, prejudice and power. Even today, years after Kuhn first presented his ideas, his view of science upsets many scientists, not in the least because he places them in the midst of the ordinary, often irrational world of human emotions, not standing apart from its *sturm und drang*, as part of a wholly logical and wholly dispassionate enterprise. But of course scientists are only human. They are not logic machines, exempt from error, unswayed by base motives, and responsive only to reason's call....

I have taught and studied biological systems at various levels of organization—from the molecular, to the cellular, to organs and tissues, and finally in whole animals—for about 40 years. In doing so, I have invariably encountered great difficulty integrating understanding at lower levels of organization into a broader perspective of the whole. I thought there were two enormous barriers. First, I could not imagine how one could explain the physiological—that is, the functional—wonders of organs and organ systems, like the heart, the circulatory system, the gastrointestinal system, the kidney, and most emphatically the central nervous system, simply by listing every last molecule and every reaction involved, no matter how well we understood those molecules and reactions. It seemed self-evident to me that this was not enough. The material instantiation of biological beings, though unmistakably crucial, was insufficient to explain physiology, and hence to explain life.

Second, and equally important, I could not see how one could discover the mechanisms that underlie various physiological processes without having first established the properties of those processes themselves, whole and intact. It seemed to be widely believed that if we could elucidate the parts of the system well enough, then sooner or later, if we had the necessary skills, insight and talents, we could understand the mechanisms that underlie any process of interest. This attitude ran counter to everything I knew about physiology. To understand the mechanism, one first had to describe the properties of the phenomenon of interest. Only then could it be grasped. This also seemed self-evident to me. How could one establish the mechanism for a particular process if that process was not known? How would we ever know what to look for?...

This problem became one of central concern in my own research in the mid-1960s. As a newly minted assistant professor at Harvard University, I had developed an interest in how cells secrete their central organic products, proteins. There was already a relatively mature model, part of a larger paradigm, that was widely believed to provide an accurate description of how this occurred. Today this model is known as the *vesicle model* or the *vesicle theory of secretion*....

When I first began to study protein secretion, I was not well versed in the evidence that supported the vesicle theory, and found it necessary to spend time in the library before I could chart my own path. I had to understand what was known, but also where the lacunae of ignorance lay. What I discovered was a considerable shock to me. Rather than being substantial, the evidence, what there was of it, seemed weak and ephemeral. And worse yet, often what was considered evidence was far more hypothesis than proof. This was not only a great surprise, but raised two worrisome questions. First, why was the vesicle model so widely perceived to describe the actual events of secretion in the absence of convincing proof? And second, given its lack of foundation, how did it come to be so specific and so complex? I anticipated that such a detailed model would be backed by far more convincing evidence than I was able to find.

Equally disconcerting, there seemed to be precious little known about the process it sought to explain. As best I could tell, the model professed to explain a process that was barely understood beyond its mere occurrence. Of course, it was known that secretion occurred, and also that it could be increased or decreased by various stimulants and inhibitors, but that seemed to be about all. This seemed something like sending a rocket to the moon, knowing that the moon existed and knowing its general location (up there and relatively far away), but not knowing how far the rocket would have to travel, in what direction, or to how large a target. It might get there with a great deal of good luck, but you certainly couldn't count on it.

It was with these realizations that my students and I began our experiments. We sought to connect the actual, measurable properties of the process of secretion to the underlying mechanism. As we explored the natural system, the vesicle model served as our guide by providing a framework for predictions. The task was to check the natural process for the properties predicted by the model. If we found them, it would be affirmed. If not, it would have to be rejected or modified.

Rather than salving our concerns, our initial tests raised more questions about the vesicle model than they answered. This spurred us to make further predictions and undertake further trials. But the additional results only served to intensify our uncertainty. It was not long before we began to publish our observations, and wonder aloud (that is, in print) about the goodness of fit of the standard vesicle model to the process it was thought to explain. If secretion did not occur as the vesicle theory proposed, then now did it occur? Along with our questioning, we introduced an alternative hypothesis for the mechanism of secretion. We can call it *direct transport* for the moment.

In reporting this research and commenting on its implications, the die was cast. Quite naively, at least in retrospect, we expected that others would see the weaknesses in the vesicle model as we did, and would join us in further attempts to validate or modify it to more accurately reflect nature. For the most part, this did not happen. Most experts in the area assumed, much to our chagrin, that our observations were erroneous or invalid, however straightforward they seemed to us. Our contradictory evidence was put in doubt, not the vesicle model. And when it eventually became clear that our observations were accurate, it was assumed that they must fit the vesicle model somehow, even if it was not clear how.

In part this was because it was assumed that the alternative we had proposed was not possible. Although this assessment has subsequently been shown to be wrong, at the time it was inconceivable that the case could be otherwise. Since direct transport was the only available alternative, and it was not believed to be a credible choice, the vesicle model had to be correct; and our results, their interpretation, or both had to be in error. This was the opening volley in a controversy that lasted for decades. ..."

From the Introduction to the Chinese translation of *Lessons from the Living Cell*:

"... The belief that we can come to fully understand the processes and mechanisms of life through the deep and comprehensive study of their constituent elements, most significantly life's central chemical substances DNA and proteins, has come to dominate thinking in most areas of biological research. Though our far-reaching mastery of life's material nature is in great part attributable to this approach, in the final analysis it cannot fulfill the expectations of its advocates. It introduces, ineluctably, two profound confusions and misapprehensions. The first is that life's material incarnation is identical to, co-extensive with life itself, and the second misidentifies the whole as the mere sum of its component parts.

... Complete reliance on this reductionist viewpoint, what I call the strong micro-reductionist principle, not only leads to these conceptual mistakes, but to a serious distortion of scientific method (in particular, abjuring the need to test scientific theories according to the principle of falsification). ... (*As the philosopher Karl Popper explained the principle of falsification is comprised of the following*): 1) the failure of scientific theories in the face of tests of their validity requires that they be rejected, 2) since no confirmatory test can forswear other tests that a particular theory might fail, all affirmative scientific understanding must be viewed as conditional, and 3) theories incapable of falsification, of being shown incorrect or false, are not proper scientific theories.

... Some 40 years ago, as an idealistic young scientist, a naïve Popperian at Harvard I would say, I found myself in the middle of a heated controversy much against my will and desire. By both training and inclination I had come to think of science as the testing of theories about nature and determining whether or not they passed muster. ... Scientists were duty bound, I thought, to discard theories that did not comport with nature's properties, however captivating they might otherwise be. This was science's Hippocratic oath. Above all it was what it meant to be a scientist.

I had just published observations that conflicted with an influential theory of cellular function, a pillar of the modern view of the cell as a mechano-chemical machine that concerned how cells secrete proteins. The reaction that this work and the experiments that soon followed received from experts in the field surprised me greatly. Rather than expressing interest and curiosity, rather than questioning their theory's validity in the face of negative results, they simply dismissed the new observations. Some were angered and literally screamed that the results were artifacts, and that even if they could be convinced they were real, they had to be explained in terms that were in harmony with established theory no matter what. That their model was incorrect was unthinkable.

Unknowingly and unwittingly, I had met the paradigm, that behemoth of scientific belief, and its battlements were being defended. My idealistic world of the falsification of tentatively held theories had run headlong into the real world of science, with its strongly held beliefs, with its paradigms. Additional studies deepened my skepticism, but just as surely they deepened the skepticism of my antagonists: mine about the paradigm, theirs about my observations. And so the controversy grew.

Outside of my family and some of my students, I felt alone, a minority of one, while supporters of the paradigm not only had the power of numbers, not only worked in or ran large and well-supported laboratories, but also controlled scientific journals and professional organizations. They were so well connected that they literally held sway or at least had great influence over whole fields of study. Nonetheless, however broad, extensive, or sophisticated their research, in neither its design nor interpretation did it test, question or otherwise challenge the model they believed in so fervently. They sought only its confirmation.

It was in this circumstance, that I first came across Thomas Kuhn's book. I found it simultaneously eerie and comforting. It was eerie because it quite accurately described what I was experiencing even though it was written before I experienced it. He seemed prescient. It was comforting because it assured me that I was not alone. Though I was only a novice, the research and views of many who became giants of science were similarly dismissed and denigrated by their peers, and in the same ways for the same sorts of reasons. This helped me not personalize my experiences and made me realize that my idealistic and idealized views of science were just that. Science was not what it claimed. Despite the frequent posturing, it was like most human endeavors, in significant part a political and social enterprise, not a simple search for the truth.

Popper had it right about how science should be practiced, but Kuhn was right about how it is practiced. Science can be an ennobling human endeavor, and I will be forever grateful for having had the opportunity to ask basic questions about nature as part of my life's work. I have been blessed with many of its riches. But however ennobling, we must not delude ourselves about science's real nature. Paradigms abound, sometimes hidden, often unrecognized, constraining and forming the research enterprise. As a consequence, we must always be on guard, prepared to distinguish what is truly known, from what is just believed, just paradigmatic."

*section one*

---

*Intimations and forebodings*

## chapter one

# At the bench
## Watching and learning

> The prudent man always studies seriously and earnestly to understand whatever he professes to understand, and not merely to persuade other people that he understands it; and though his talents may not always be very brilliant, they are always perfectly genuine. He neither endeavours to impose upon you by the cunning devices of an artful impostor, nor by the arrogant airs of an assuming pedant, nor by the confident assertions of a superficial and impudent pretender. He is not ostentatious even of the abilities which he really possesses. His conversation is simple and modest, and he is averse to all the quackish arts by which other people so frequently thrust themselves into public notice and reputation.
>
> **Adam Smith**
> *The Theory of Moral Sentiments, 1759*

My father was an alchemist, or so it seemed to this five-year old who had just been given permission to watch him perform his magical transformations. I sat quietly, transfixed, on a stool across from the high bench, where he carried out all sorts of strange acts. To manipulate his materials, he employed strange-looking implements with foreign-sounding names such as *spatula, mortar,* and *pestle.* To apportion things correctly, he used what seemed to be measuring devices, glass cones for fluids and balances for solids. With these and other tools of his trade, he created, material-by-material, method-by-method, elixirs, syrups, unguents, powders, ointments, creams, capsules, suppositories, and a variety of other unnamed things.

Though I found what he was doing spellbinding, my father was not actually engaged in supernatural activities. The bench on which he plied his craft was in the back of his small pharmacy, *Rothman's,* on the corner of Gun Hill Road and Hull Avenue in the north Bronx, up the hill from

Bronx park and a block away from Woodlawn Cemetery with its notable denizens. It was one shop among many on Gun Hill Road that served this small, largely Jewish neighborhood and the adjoining Irish one stacked with five- and six-story apartment buildings mixed in with single-family dwellings for the more fortunate.

My brother and I grew up in this neighborhood and in equal measure in our father's drug store. Like so many children of his time, my father had worked since he was nine years old to help support his impoverished family. One of his jobs was a stock and delivery boy in a pharmacy. Eventually, he was able to put enough money aside from his various jobs to pay tuition at Fordham University, where he studied to become a medicinal chemist: a pharmacist.

However otherworldly his actions seemed to me, what he was doing was based on the quite worldly physical and chemical characteristics of the substances he was manipulating with the object of creating medications for his customers. Though a revolution in drug manufacture was underway – making the production of drugs a factory operation carried out by machines, not pharmacists – at the time, pharmacists still performed many of these procedures manually in the chemical conjuring they called "compounding."

Over the years as I watched my father ply his craft, I learned that he did his work with exceptional skill, accuracy, and great attention to detail, checking everything and then checking it again. Each item was measured and weighed with extraordinary care, put together in a prescribed sequence, and often at a specified rate, in accordance with a formula based on the physical and chemical properties of the substances he was manipulating. Was the material soluble or insoluble, in what fluid, in what amounts, and at what temperature? Were fluids miscible or would they separate? And would X react with Y and was this to be encouraged or avoided?

To my knowledge, in over 30 years of compounding prescriptions, nothing left the pharmacy that was not as it should have been. There could be no error. Too much was at stake – other people's lives of course, but also our family's livelihood. An error could lead to someone's death or to grievous bodily harm, and by law or rumor, it could end my father's career.

His singular insistence on doing it right, carefully, and accurately became part of me. Whatever I was to do in life, the ways in which he carried out his magic at the bench would not only follow me, but at times would haunt me. He was the alter ego sitting on my shoulder in the laboratory. *Stephen, be careful, pay attention to what you are doing, be observant!*

## chapter two

# An indifferent student
## Of clams and starfish

> I would rather have questions that can't be answered
> than answers that can't be questioned.
>
> **Richard Feynman**

Late one night, I found myself in the zoology building attempting to memorize the names and relationships of what seemed an endless list of species for a final exam on invertebrate zoology. I took to wandering the halls to avoid this unpleasant task. In one of my peregrinations, I saw light coming from a small lecture room down the hallway. I walked to it and peered through the open door. What I saw was very odd.

A handful of students, scattered in the seats, were looking with rapt attention at a long, high laboratory bench at the front of the room. A demonstration of a sort was underway. On top of the bench crouched on all fours was old man, a Professor Heilbrunn, attempting as best he could to look like a cat. He was trying to demonstrate to the perplexed students how cats walk. Being human, this was no easy task. Nonetheless, using his body as the model, he explained with great enthusiasm the differences between the anatomy of human and feline limbs. Professors were usually long gone by this time, and as it turned out he was not an instructor of comparative anatomy, yet there he was teaching it in this peculiar, but very effective way.

I decided then and there to take a course, any course that he taught. His specialty was general physiology, the study of the fundamental properties of living things, and befitting a man crouching on all fours on top of a laboratory bench, his course was unusual for the time. Not only were the students asked to perform a series of standard cookbook laboratory experiments, but they also had to carry out experiments of their own design.

About halfway through the semester, he invited the class to his little house on Smedley Street, a tiny Bohemian enclave in Philly, all of one block long, to drink beer and design experiments. The experiments could be about almost anything, though he offered a list of open questions for we clueless students to contemplate.

Along with my equally oblivious partner, we chose from his list. The question was, how did the starfish open the clamshell to get at the tasty morsel inside? We were attracted to the project, not so much out of intellectual curiosity, but because we would have a ready supply of clams to steam for dinner. Now that was motivation!

The view, common at the time, was that the starfish did the dirty deed as a result of suction applied by the many small tube feet on its ventral or underside that stretched out from its mouth at the center to the tips of its arms. It has since been learned that the tube feet secrete an adhesive to achieve attachment and use muscles to retain it. In any event, the tube feet were thought to be responsible for holding starfish to rocks in the face of fierce tidal currents and so it seemed likely that the same mechanism was used to pull the two halves of the clamshell apart.

But there was a problem with this idea. Though it requires some effort, it is possible to pull a starfish off of a rock, but try opening a closed clamshell without a tool to pry it open. It is futile. If we can pull starfish off rocks, but are not strong enough to open clamshells, were tube feet up to the job? Our first task in attempting to answer this question was quantitative. It was to measure the force required to open the clam despite the best efforts of its powerful adductor muscle to keep it closed. Was this really beyond the capability of the tube feet?

Towards this end, we drilled holes in both sides of the clamshell near the opening and attached cables. Then with a simple lever we applied graduated forces of equal magnitude to both sides of the shell measuring the width of the opening they produced. When we plotted the applied force against width of the opening, we found two distinct phases. First, a relatively modest force opened a slight crack, less than a millimeter wide, between the two shells. The second phase required far more force, but with enough, the muscle gave up and the clam was exposed. But the tube feet of the starfish seemed to be weak tea pitted against the powerful adductor muscle. If this was correct, then how, crouched over its prey, did the starfish open the clam? It seemed to have no other means.

We realized that we had found the chink in the clam's formidable armor. It was the first phase, the small opening produced by a slight force. We thought that it was through this opening that the starfish defeated the clam's powerful muscular defenses. We imagined a poison of some sort being secreted by the starfish into this space to incapacitate the adductor muscle.

To test this hypothesis, we needed to obtain the poison, and for that we needed its source, the starfish. And so we found ourselves visiting tide pools near the Jersey Shore where starfish were plentiful. Since there were no laws at the time to admonish us, we collected enough starfish to extract the presumptive poison. We then removed and ground up glands located in each arm, thinking that they were the most likely source of the poison. We then injected a small amount of the extract into the space and waited for the shell to open. But nothing happened. The clam remained resolutely shut, clammed up.

Our experiment had failed. We found no poison. But, as it turned out, we were on the right track. Among the glands we had ground up were the digestive glands, and the enzymes they secreted were the sought-after poison. They destroyed the powerful muscle by digesting it. Either we had injected too little or were not patient enough. The starfish's trick was that rather than injecting the enzymes as we did, it everted its extremely thin stomach, inserting it directly into the clam through the tiny crack. Once inside, the stomach secreted enzymes for as long as needed to digest the clam, muscle and all. Great force was not needed, just the Trojan horse of a stomach that could fit through the tiny crack.

Though we had not figured this out, I was nonetheless very excited by the experiment. It seemed magical. I had asked a question of nature, and nature had responded. I had learned something about the properties of clams and starfish. I knew that scientists asked questions of nature and obtained answers all the time, but that this indifferent student from the wilds of the Bronx could ask them, and nature in its grace would provide answers, was electrifying. Though I had no idea how I would do it, I knew then that asking such questions and seeking answers was if possible going to be my life's work. The path turned out to be long and tortuous, not to mention arduous, but I persevered and, for both better and worse, became a research biologist able to ask fundamental questions about life's properties and seek answers. And wonder of wonders, time and again, nature answered![1]

## chapter three

# The dark in the window

## The real thing

> The first scientific postulate is the objectivity of
> nature: nature does not have any intention or goal.

**Attributed to Jacques Monod**

It was dark. I was cold, exhausted, getting nowhere, and feeling adrift. It seemed that the joy of doing scientific research had evaporated long ago as I faced the real challenge of actually identifying a problem for my thesis and tackling it. Mesmerized, I was looking out of the two large horizontal swathes of glass that formed the two adjoining outward facing walls of the room in the striking brick and glass tower that was, I had been told, a great architectural masterpiece. To me it was a chilly and indifferent cinderblock warehouse. That evening, in place of the usual view of the old dormitories and the tree-laden walk below, there was a faint reflection of the contents of the room in the blackness of the windows—two ugly government-issue green metal desks and an equally ugly green metal rectangular table, along with four barely serviceable metal chairs. The room was an office that I shared with my supervisor and that also served as a makeshift laboratory for me.

On the table, laid out on its back, was an anesthetized rabbit, its white fur torn apart at the midline by a long incision covered by a few wet gauze pads to prevent drying. Poking out between the gauze pads was a fine catheter that wended its way to a small glass test tube sitting on the table below, slowing filling with a clear fluid from the animal's pancreas.

Unlike the human pancreas, the rabbit's pancreas is diffuse, a thin, wispy piece of tissue traversed by a single large duct, called without fanfare the pancreatic duct, into which many tributaries feed as it passes through the gland to the intestinal lumen. The catheter had been placed in the duct through the small orifice on the interior or mucosal surface of the intestinal wall and was tied in place at its external or serosal surface.

I turned back into the room, abruptly ending my anxiety-riven reverie, picked up a syringe, filled it with a lethal dose of anesthetic, and injected it into the rabbit's femoral (leg) vein. When the rabbit stopped breathing, I removed the gland with a pair of surgical scissors along with the first segment of intestine, the duodenal loop, in whose omentum (connective tissue) it was embedded. With the catheter still fixed in place, I placed it all in a large Petri dish filled with a physiological salt solution to keep the tissue wet.

My plan was to cut out and chemically fix small pieces of tissue for future examination under the microscope. But before I was able to do this, I noticed fluid dripping from the open end of the catheter onto the table. I waited, expecting it to stop at any moment, but it continued. Seeing this, I spread the tissue out in the dish, began gassing the fluid bathing it with oxygen, and put a new vial in place to collect the secretion. Remarkably, under these far from ideal circumstances, secretion from the spread-eagle gland continued undiminished for close to 2 h. Even more remarkably, its rate was roughly the same as it had been in the animal *with the gland's blood supply intact* (Figures 3.1 and 3.2).

The gland, it turned out, was able to secrete fluid in the absence of blood pressure and blood flow. This discovery was of both practical and theoretical significance. As for its practical significance, this meant that the rabbit pancreas, whole and intact, could function outside the animal, that is, *in vitro*, and therefore that fluid secretion by the gland and its directional nature could be studied away from the complexities of the animal, including those introduced by blood pressure itself. Secretion could be examined in simple and stable surroundings. There would not be any nervous or hormonal input or other extraneous, variability-inducing influences of unknown origin to confuse matters.

In addition, I would be able to control the gland's environment. Most importantly, I would be able to control the character of the medium in

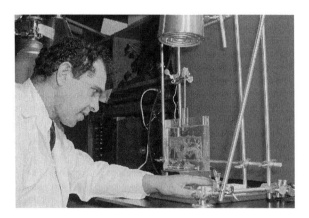

*Figure 3.1* The author in 1962 as a young graduate student sitting at a table next to a live rabbit pancreas in an incubation chamber.

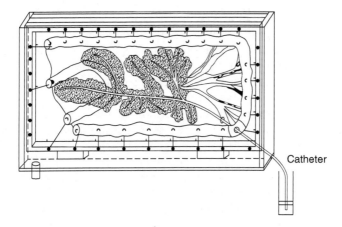

**Figure 3.2** An early drawing of a live rabbit pancreas in an incubation chamber. The pancreas is suspended in the connective tissue (omentum) of the first loop of small intestine (the duodenal loop). The loop serves as scaffolding for the pancreas and in turn is held in place by stainless steel hooks attached to a plastic (polymethylmethacryate (PMMA)) frame mounted in the chamber. "Catheter" refers to the polyethylene tubing fixed in the duct that carries fluid secretion from the gland into a vial.

which it was bathed. I could add a substance here or delete one there, or change its concentration. I could then determine how the changes I made affected secretion. From such knowledge, I hoped to decipher its mechanism. This thin, translucent, almost invisible gland had given me the opportunity to expose the deep secrets of the mechanism of secretion. All I had to do, or so I thought, was devise and construct a chamber to house the gland, and carry out experiments designed to discover the mechanism.

The *theoretical* significance of what I had chanced upon concerned a belief that dated back many hundreds of years to the discovery of the blood circulation and glandular secretion themselves. Whenever and wherever fluid is secreted, there is a flow of water across the secreting tissues or organs, across the secreting surface, from blood to some final destination, usually external compartments such as the gut or the surface of the skin. The fact that blood is the immediate source of this water led to the sensible conclusion that the motive or driving force for its movement, for secretion, was blood pressure. That is, it seemed that secretion was the result of the pressure ultimately produced by contraction of that great muscle, the heart.

But I had just observed secretion in the absence of blood pressure. This was proof that, to the contrary, blood pressure was *not* the source of energy for the movement of water across the pancreas. It apparently came from elsewhere. And if this was so for the pancreas, then it might be so for secretion in general, for the secretion of water from the liver, stomach, salivary glands, sweat glands, tear glands, and so forth.

But if not blood pressure, then what produced fluid secretion? There was only one other possible source of the energy needed to move water. That was the single contiguous layer of cells that formed the secreting surface separating blood from the water's destination itself. These epithelial or surface cells serve to divide the inside of the animal from the outside (the lumen of the gut being outside the animal in this sense). They form the barrier that protects life from the outside world while they are the major means of interaction with it.

A general mechanism for how water moves across this barrier had already been proposed. The frog's skin provides a wonderful example. Most frogs live in freshwater ponds or lakes. We call them "fresh" because they contain very few dissolved substances, and in particular, very little salt. As such, they have a low osmolarity. Osmolarity is the concentration of chemical particles dissolved in a bulk solution. The osmolarity of seawater is relatively high primarily due to the presence of salt, sodium chloride, while that of pond water is quite low due its absence.

Like virtually all animals and plants, frogs are salty creatures. Their saltiness is thought to be a remnant of the seawater from whence they arose eons ago, but it is far more than that. The fresh water in which salty frogs live poses an existential threat to them as it does to all freshwater amphibians. This is because physical law seeks to dissipate osmotic gradients like that between the salty frog and the salt-free pond water. It seeks to make them equal, the same; in the language of science, it seeks to achieve equilibrium. Embodying nature's desire, pond water would enter the frog down the osmotic gradient, diluting it. This made pond water the frog's mortal enemy. If this process, called osmosis, were left unopposed, the frog would swell with water until it literally exploded.

But of course, the frog lives happily in fresh water. Indeed, this is its natural habitat. But how can that be? How can the frog protect itself from dilution and ultimate dissolution or disintegration by the inrushing water? The mechanism thought to explain the frog's good fortune is more or less the same one considered to be responsible for the secretion of water across all epithelial surfaces, including the pancreas.

Nature offered the frog three options to deal with its dire osmotic circumstances. First, its skin could be impermeable to water like an oilcloth or Gore-Tex raincoat. In this view, osmotic gradient or not, water simply could not enter. But it turned out that water could cross the frog's skin and relatively easily. So, a raincoat was not the solution. The second option was to bail out the water after it enters. Water pumps, i.e., miniature hearts or sump pumps, embedded in or between the cells of the frog's skin might pump the water back into the pond. But such pumps do not exist in the frog's skin or elsewhere in living things, and even if they did, cells do not possess enough energy to move the masses of water required.

This leaves the third option. Unfortunately, it not only doesn't help matters; it makes them worse. It requires the impossible. To get rid of the water, the frog would have to reverse the osmotic gradient that caused it to enter in the first place. In other words, the mechanism would have *to make the pond saltier than the frog*. But how could a little frog, even many little frogs, do this? How could they fill the enormous salt-free reservoir of the pond with enough salt to make it saltier?

Fortunately for frogs, there was a wrinkle in the third choice. There was a sneaky way to achieve the impossible. The frog could get the job done without having to actually perform the Herculean task of reversing the gradient between itself and pond water. All it had to do was produce a microscopically thin layer of water just outside the cell membrane in which the osmolarity was higher than in the cell. Water would then leave the cell to enter this tiny unperturbed or unstirred hyperosmotic space just outside the membrane border. From there it could enter pond water. The frog could not make the pond salty, but it could make this tiny environment, this placid microscopic ocean, salty. This was all that was needed to protect the frog from the encroaching water. It made the impossible, possible. In this way the frog could bail out the water.

Nature's solution, as it so often is, was both simple and ingenious. The cells of the skin provided this tiny environment of high osmolarity by "pumping" small quantities of salt into it. A membrane-embedded transporter with energy provided by life's premiere energy-containing molecule, adenosine triphosphate (ATP), carried out the task. Though by no means a cheap process—it has been estimated to require between 20% and 50% of the cell's energy—there is enough to get the job done, thereby allowing for the existence of the fresh water frog. Pump sodium, not water.

I expected pancreatic tissue to behave in much the same way, with an osmotic gradient driving secretion, and this became the subject of my thesis. Though I did not provide a definitive answer for how this process occurred in the pancreas, the work went well and I was even able to publish a paper describing the method and the observation of fluid secretion in the absence of blood pressure in the prestigious *Nature* magazine while I was still a graduate student.[2]

Anyway, seeing a path forward, I worked hard day and night and completed the research in about 2 years. Like all Ph.D. students, I faced one final hurdle before being awarded my doctorate. I had to write my thesis and defend it before a committee of faculty. The committee was

usually small. Mine was composed of my supervisor, an external reader—
a faculty member from another department—and another member of the
faculty in my department who knew something about the subject or who
had agreed to learn about it.

This is why, when the day came to defend my thesis, I had a great
shock. When I opened the door to the seminar room in which the defense
was to occur, there was a long rectangular table around which sat more
than a dozen members of the faculty. Why were they there? What did this
mean? It certainly could not be a good sign. It looked more like a firing
squad than an examining committee. I was terrified. My dream of becoming a scientist was about to come crashing to an ignominious end.

Reinforcing my overwhelming apprehension, at one end of the long
table sat the department chair, not my supervisor. He smiled and pointed
to the other end of the table, indicating that I should sit there. I had prepared a brief, 15-minute summary of the thesis with accompanying slides
as was customary. After the introductions, the chairman asked me to give
my talk. The audience listened attentively as I laid things out, methods
and results, and finally what I thought it all meant. They interrupted me
from time to time with probing, but generally polite questions, most of
which, much to my amazement, I was able to answer to their apparent
satisfaction.

Things seemed to going surprisingly well when I noticed that an older
professor who had been silent seemed quite agitated. He was fingering a
large black bound notebook on the table in front of him. As my talk and
the questioning came to an end, he cleared his throat and began to speak,
not to ask a question, but to issue a condemnation. He importuned in
what I took to be a British accent, "You should be careful what you claim.
On page 10, you say that yours is the first demonstration of secretion from
a whole ducted gland *in vitro* in the absence of a blood circulation. But that
is not true. You should be more circumspect."

With a flourish, he picked up the black notebook and waved it about.
"I demonstrated the same thing many years ago, not in rabbits, but in the
South African toad." I apologized and said that I would of course cite
his work in all future publications. But this did not seem to satisfy him. In
his mind, my error had put the whole thesis into question. If I had gotten
this wrong, if I was sloppy here, what else did I get wrong?

As I was struggling to satisfy him, another professor piped up.
"Where," he asked, "did you publish your results?" He responded, "I have
not published them, but they are here," pointing to the notebook he had
been brandishing about. Someone else then asked to see the notebook, to see
the data. This is when things really got interesting. He refused. Nor, as he
was pressed, was he even willing to describe the experiments he performed
or the results he obtained. As the back and forth continued, there was

increasing skepticism about what he had in fact done, and what was really in his secret notebook? At some point, it dawned on me that I was experiencing a miracle. My trial was over, my Ph.D. was guaranteed. My thesis was no longer the subject of discussion, and despite the absence of the phantom attribution, it was accepted! He never did show anyone the experiments.

It was not my research that had upset him. His concern, apparently his sole concern, was his priority. He wanted the credit that he thought he was due. He was angry that I had not acknowledged his discovery. Though I did not know him, the issue was personal, not scientific. To this day, I am certain that if it had not been for the intervention of a few good Samaritans, my thesis would have been rejected.

This was my introduction to the often contentious and unpleasant side of science, and it was a premonition of a future filled with conflict. Though I had no idea at the time that my career was going to be charged with controversy and acrimony. Before long, as explained, and much against my desire, inclinations, and expectations, I found myself in a bitter conflict with a whole field of biology about an enormously consequential model for which, it turned out, a Nobel Prize was about to be awarded.

That model concerned secretion by the self-same pancreas, not of fluid, but of protein, in particular the digestive enzymes. In mammals, the pancreas accounts for the manufacture and secretion of most of the enzymes responsible for the digestion of food in the intestines. It is a premiere protein factory, perhaps the most prodigious source of protein in the body gram for gram. The model in question explained how newly manufactured digestive enzymes were moved from place to place within the pancreatic cell and then secreted.

The problem I faced was not that I had once again ignored the results or ideas of others. Quite to the contrary, the experiments I performed were based on them. The issue was one of compliance. My opponents knew to a certainty that their model, by then the standard model, was correct, and consequently, that my results, observations that had raised questions about its dominion, had to be incorrect. All they wanted from me was to admit error, to admit that my results were false, artifacts. As with the professor with the black notebook, they were not interested in my experiments. They wanted them gone, excoriated, erased. What they wanted was compliance, my compliance with their theory.

In any event, unaware of the inharmonious future that awaited me, I received my doctorate that spring, looking forward with great optimism and keen delight to an exciting career of discovery as a scientist. I spent the next year finishing and publishing my thesis research,[3,4] while looking, more like hoping, for a job at a university that came with a laboratory in which I could to carry out my next scientific project. It seemed that a wonderful and satisfying career awaited me, if only I could find a job.

I did, and it was at Harvard. I had been given a great opportunity, and yet the die was about to be cast and my unkind fate determined. Like it or not, my future as a scientist was to be fraught with anger, rant, and attack directed at me. It all began innocently enough with a story that my thesis supervisor, Frank Brooks, told me about an old controversy that involved the famed physiologist Ivan Pavlov. It turned out that I was well positioned to resolve the controversy using the preparation of rabbit pancreas I had developed for my thesis.

We agreed that in the few months before I moved on, I would resolve it. As things happened, not only did I not resolve the old controversy, my research ended up breathing new life into it. More than that, the dispute reached well beyond Pavlov's issue, to critical questions about the fundamental nature of the cell. The technically and theoretically simple experiments I was about to carry out were to set the course of my career for much of the next half century. By now, the controversy they engendered is an old friend whose story I am about to share with you all these years later.

## chapter four

# The observant Dr. Pavlov
## Digestion, saliva, and hormones

> It is a piece of idle sentimentality that truth, merely as truth, has any inherent power denied to error, of prevailing against the dungeon and the stake. Men are not more zealous for truth than they often are for error, and a sufficient application of legal or even of social penalties will generally succeed in stopping the propagation of either. The real advantage which truth has, consists in this, that when an opinion is true, it may be extinguished once, twice, or many times, but in the course of ages there will generally be found persons to rediscover it, until some one of its reappearances falls on a time when from favourable circumstances it escapes persecution until it has made such head as to withstand all subsequent attempts to suppress it.
>
> **John Stuart Mill**
> *On Liberty, p. 37*

It is said that it was Ivan Pavlov's habit to walk unannounced through his laboratory to check on students, fellows, and technicians as they were carrying out experiments. On one particular day as he was making his rounds, something caught his eye that has become as famous as Pavlov himself. In a classic example of serendipity, the gift for discovery, he noticed something out of the ordinary about a dog with a salivary fistula, a gland he had surgically modified so that instead of secreting saliva into the mouth, fluid exited the duct onto its cheek.

Whenever a certain technician walked past the animal, he began to drip saliva. Pavlov asked about the relationship, if any, between the technician and the dog. As it happened, the technician was responsible for feeding him. Pavlov realized that the dog salivated in the technician's presence because he was *expecting* food and stopped secreting when he left because his expectations were dashed. He was not salivating because food was in his mouth, the process Pavlov had been studying, but in anticipation of eating. He was predicting this desirable future and reacted accordingly.

For most people, scientist and layperson alike, things would have ended there. After all, we all salivate when we are about to eat something delicious. But what it brought to mind for Pavlov would forever change our understanding of human and animal behavior. He recognized that he was observing a previously undiscovered reflex.[5]

At the time, reflexes were understood to be automatic. A particular experience would elicit an involuntary, unthinking mechanical response from the animal, for instance, food in the mouth or stomach would cause secretion, or pressure on the skin would cause a muscle to contract. Its automaticity is what it meant to call something a reflex.

What Pavlov had observed was a different kind of reflex, a learned reflex. There was no food in the dog's mouth and yet he secreted saliva. The dog seemed to be thinking that the technician's presence meant that he would be fed shortly. The dog had made an association between the technician and food, and was secreting saliva in anticipation of eating.

Pavlov called automatic reflexes, unconditioned because no prior experience was needed to elicit them. The presence of the stimulus was both necessary *and sufficient* to obtain the response. He called his new reflex a conditioned reflex because the presence of the stimulus, the technician, was not in itself sufficient to produce a response. That required learning, habituation, in other words conditioning. The dog had taught himself to secrete.

Using his famous bell in place of the technician, Pavlov explored the nature of the conditioned response. In his initial studies, he paired the ringing of the bell with the offer of food. After some experience, after a number of trials pairing the two, Pavlov uncoupled them. The bell was rung, but there was no food. In its time, what he saw was so strange, that it seemed supernatural. The dog secreted saliva to the sound of a bell, just the sound of a bell. The bell had become a sufficient stimulus.

How could that be? How could something totally unrelated to its purpose cause salivation? Over time, the withholding of food led to a decrease in the response, and it eventually disappeared. The sounding of the bell was no longer a sufficient stimulus for salivation. The response was "extinguished." The dog had learned the opposite. He came to realize that the ringing bell no longer meant that food was on the way and stopped secreting. First, he had learned to respond and then not to respond.

Pavlov understood that what he was observing was not just about food and salivary secretion, but was a general phenomenon. Responses of all kinds could be learned, and so the field of behavioral psychology or behaviorism came into being, and although he was not its sole creator, his conditioned reflex was its avatar. Just as the dog responded to the bell, animals and humans alike could be taught to behave in all kinds of ways if the stimulus for the behavior was reinforced. Though food was the most basic means of conditioning, of positive reinforcement, there were other effective inducements, other kinds of comfort and soothing. Learning could also take place using negative incentives, negative reinforcement. For example, inflicting discomfort, cold, hunger, or pain could also produce a desired response.

Pavlov's discovery was serendipitous in another way. It was the unanticipated consequence of his research over the prior 20 or so years on the regulation of digestive function by the nervous system. This was what his associates were up to on that historic day. They were studying the regulation of salivary secretion by the nervous system. His research on the regulation of gastrointestinal or digestive function had been seminal. It had laid the groundwork for our modern understanding of these processes, and he was awarded the Nobel Prize for this research in 1902.

The central premise of this work was that the digestion of food was regulated by the nervous system. This was hardly a revolutionary idea, since at the time the nervous system was the only known means of regulating bodily functions, and of course, digestion was such a function. But as fate would have it, within a few years of Pavlov receiving his Nobel Prize, a discovery made in London changed everything.

Two physiologists, Ernest Starling and his brother-in-law William Bayliss, discovered a means of biological regulation that did *not* involve the nervous system. They found that a substance made in one tissue could affect the action of another remote tissue by passing through the bloodstream. Like Pavlov, they were studying the gastrointestinal system. They had extracted a substance from the duodenum of one dog that produced fluid secretion from the pancreas of another dog after being injected into its venous circulation.

They called it "secretin" for obvious reasons. A few years later, Bayliss coined the term "hormone" to describe secretin and any and all substances that like it were made in one tissue and acted on another remote tissue after having been released into the bloodstream. They had discovered a whole new world of biological regulation—hormonal regulation—and with its discovery, with the discovery of hormones, a new field, endocrinology, came into being.

It did not take long before Pavlov, back in St. Petersburg, heard about these new experiments. He was skeptical, if not downright dismissive. He wrote Starling and asked if he would be willing to demonstrate the effect for him if he came to London, expecting him to put him off. Much to Pavlov's surprise, Starling agreed. And so he found himself taking the arduous trip to London sure that Starling and Bayliss were up to no good.

When the day came for the experiment, he watched attentively as they went about their business. First the duodenum of an anesthetized dog was excised. Next, it was cut into small pieces, placed in a mortar along with sand and saline, and ground up with a pestle. This resulted in a cloudy suspension that they then filtered through gauze to remove the sand. Without a fare-thee-well, the sand-free material was then injected into the venous circulation of another anesthetized dog whose pancreatic duct had been fitted with a catheter to collect secretion.

All this must have seemed surreal, even insane to Pavlov. What could one learn from the injection of this cloudy mess of ground-up tissue into the bloodstream? That it would cause the pancreas to secrete fluid seemed far-fetched. Given his negative predilection, what happened must have amazed him. Soon after the injection, fluid began dripping out of the catheter. Drops soon turned into a stream, and then into a virtual torrent as fluid exited the pancreatic duct. Obviously, there was something in that mess. It took over 50 years to find out what it was, but eventually it was discovered that secretin was a peptide, an amino acid chain, 27 amino acids long, that was a fragment of a larger protein.

Pavlov returned to St. Petersburg disturbed and perplexed by it all. It is said that he locked himself in his office, only emerging after several days to call a meeting of his staff. At that meeting, he made a momentous announcement. His laboratory would cease working on the digestive system. It would henceforth devote itself solely to the study of the conditioned reflex. He would have no part of this newfangled biology, with its hormones.

Starling's messy suspension had made an absolute mess of Pavlov's sense of an all-controlling nervous system. And so it was that the study of the conditioned reflex occupied the remainder of his career. This was sad despite the importance of the conditioned reflex, and not merely because he had abandoned one important area to focus on another. It was sad, indeed, it was heartbreaking, because his work on the conditioned reflex became a tool of oppression as the socialist state sought to control human behavior applying his ideas about conditioning to achieve social and political goals.

Forgotten amidst all this was arguably Pavlov's greatest contribution. In the late 19th century, the study of nervous reflexes was a central pursuit of many physiologists. Unlike Pavlov, their interests mainly concerned the body's mechanical properties, for instance, the regulation of cardiac and respiratory function, the relationship between sensory stimuli such as touch and the motor (muscle) responses they produced, or the internal regulation of muscle tension.

But Pavlov's interest was in the digestion of food by animals, and though muscles were involved in this process and so it had a mechanical element, it had recently been determined that digestion was primarily a chemical property. Though biological chemistry was still in its infancy, against centuries of belief, the great French physiologist Claude Bernard along with the German histologist Rudolf Heidenhain and a variety of others discovered that the digestion of food was chemical, not mechanical. Molecules called enzymes, whose character was still unknown (today we know them as proteins), bore primary responsibility for the breakdown of food, not, as had been thought for millennia, mechanical grinding.

As for Pavlov's forgotten but magnificent contribution, it was that all chemical reactions in living things, unlike chemical reactions outside them, were regulated. He said that for life to succeed, for it to be stable, the rates of its chemical processes had to be kept under tight control. Today, the study of the regulation of biochemical reactions plays a vital role in virtually every field of experimental biology, and is arguably as important as research on the reactions themselves.

Given his interest in the gastrointestinal system, the regulation of digestive reactions was naturally foremost on his mind. In the absence of other options (i.e., before the discovery of hormones), Pavlov assumed that their regulation was the job of the nervous system. As he imagined it, the rates of secretion of different digestive enzymes were altered by the nervous system to suit the needs of the meal undergoing digestion. If a meal was comprised primarily of meat, then enzymes that broke down meat would be preferentially secreted; if it was bread, then enzymes that digested bread would be secreted, and so forth.

The intestine was the primary site of these reactions, and the pancreas was the major source of the enzymes responsible for them. Due to his stature, Pavlov's ideas about the regulation of digestion were accepted without hesitation, even though in point of fact they were merely notional. In other words, they were supported by little if any evidence. Proper measurements of the enzymes could not be made at the time, and in their absence, his ideas could not be appropriately assessed.

Nonetheless, his point of view held sway for many years, until one of his students, B.P. Babkin, had the impertinence to challenge the master. Against Pavlov, Babkin argued that the pancreas secreted all of its enzymes at a single rate. They were secreted *en masse*, all of them together, whatever the circumstances, whatever the contents of the meal, not at different rates. There was *no* regulation of the reactions of digestion, except with regard to the amount secreted. There might be more or less secretion, there might be more or less digestion depending on the size of the meal, but its contents, what was being digested, was irrelevant to the mix of enzymes the pancreas secreted (Figure 4.1).

Though proof for Babkin's point of view was hardly overwhelming, unlike Pavlov, he had some evidence. Though still primitive, relatively

**Figure 4.1** Ivan Pavlov (on the left) and Boris Babkin (on the right). In the middle is Professor Anrep from Cambridge who translated "The Conditioned Reflex" into English.

inaccurate and nonspecific, the rates of secretion of some pancreatic enzymes could now be estimated. Using then-available methods, Babkin assessed the relative rates of secretion of several enzymes in response to meals and pharmacologic stimuli. From these measurements, he concluded that their proportions in secretion remained *constant*, unchanging, whatever the stimulant. The pancreas secreted the same mixture of enzymes, whatever one ate. The proportions of the different enzymes, known today to number about 20, was invariant. The body, he claimed, does *not* vary what it secretes to comply with the particular needs of the meal undergoing digestion. He called this "parallel" secretion.[6]

Given the serious shortcomings of his studies, the fact that everything looked the same, looked fixed, was not surprising. The values he obtained varied widely from animal to animal and trial to trial, and he only reported a few isolated experiments. Most troubling, the methods used to measure the enzymes were insufficiently specific and inadequately controlled to make clear determinations of their levels. With the uncertainty, sparsity, and importantly the variability of his observations, even the most sophisticated statistical analysis might not have exposed differences in the rates of secretion of the different enzymes, even if they were prominent.

One could even argue that, given these circumstances, his conclusion was inevitable. The deficiencies of his measurements were allies of the null hypothesis, the conclusion that there is no difference. As a general matter, if you believe that differences do not exist between two conditions, such deficiencies are your dearest ally, your best friend. It is easy to find nothing, even when there is something to be found. Whether inherent, methodological, or simply due to sloppy technique, if the variability is great enough, if the number of observations are few enough and if the measurements are uncertain enough, then whatever differences may exist in nature will simply not be seen.

It was under these far from ideal circumstances that Babkin had concluded that differences do not exist. When you got down to it, this was not much of an improvement over Pavlov's evidence-free notion. Still, Pavlov was out of the game, isolated in the Soviet Union studying the conditioned reflex, while Babkin was free in Canada at McGill University. As a consequence, he replaced Pavlov as the consummate authority on the subject, and "parallel secretion" became the standard understanding. So much so that it became common practice to only measure one enzyme in secretion, most often the starch splitting enzyme amylase, because one could safely assume that everything else followed suit. It was understood that amylase, indeed any pancreatic enzyme, could serve as a proper doppelganger for them all.

And so it came to be understood that the proportions of enzymes in pancreatic secretion were fixed and unresponsive to changes in the contents of the meal, the needs of digestion otherwise, or the nature of the stimulant. This old controversy between Pavlov and Babkin intrigued me. Pavlov's idea that digestive reactions were regulated made good *prima facie* sense despite the lack of evidence. Why would the body leave these reactions, so critical to animal life, unregulated or only regulated en masse, that is, without specificity, when all other biochemical reactions seemed to be regulated with great, one could even say with supreme specificity?

My interest in the controversy was stimulated by the fact that I thought I was in a good position to resolve it. The organ culture of pancreas I had developed for my thesis made it possible to measure the secretion of digestive enzymes in a far more controlled setting than was possible in the animal, in situ, where variability due to unknown and uncontrollable influences was great. Equally important, it had become possible to make accurate and specific measurements of at least some pancreatic enzymes.

As noted, I decided to use my remaining time at the university to resolve what seemed to me to be a relatively straightforward question. I would test the opposing points of view and clarify the situation. I intentionally selected circumstances that were unlikely to confirm Pavlov's position. To my way of thinking, if his ideas were confirmed under severe and limiting conditions, it would be especially convincing. Towards this end, I compared the secretion of two enzymes that were close cousins, trypsin and chymotrypsin. Not only were they of essentially the same molecular weight and same shape, some 50% of their amino acids were homologous, that is, were at the same location along the peptide chain. Moreover, both molecules were proteolytic or protein-splitting enzymes and hence were concerned with the digestion of the same foodstuff. More than that, both acted in the same way, facilitating the splitting of bonds within the peptide chain, not as some other enzymes did at its ends.

Yet despite all their similarities, the chemical activities of the two enzymes were completely different. Trypsin splits the chain at bonds that contain basic amino acids (producing basic solutions), while chymotrypsin is active at sites that contain aromatic amino acids (those with ring structures). Along with this strict specificity of action, synthetic substrates had recently been developed that were equally specific and that made it possible to distinguish the relative presence of these two very similar molecules with some precision. Were their proportions constant, or did they vary? Who was right, Babkin or Pavlov?

As a stimulant, I chose a hormone from the duodenum, called pancreozymin, which was thought to increase the secretion of *all* pancreatic enzymes. Oddly, this substance is known today as cholecystokinin (CCK), a hormone that causes the gall bladder to contract. This peculiar nomenclature was the result of a failed effort over many years to separate the two activities from each other. Eventually, there was acquiescence. The same molecule produced these two very different effects. Today, we know that CCK has a variety of actions, including an important role in the nervous system. In any event, CCK was settled on as the name for the hormone, simply because its action on the gall bladder was discovered first.

I thought that my preparation of rabbit pancreas, the choice of enzymes, and the choice of stimulant provided a demanding test for Pavlov's point of view. If no change in their proportions was observed, it would support, though not certify Babkin's position. If, on the other hand, the proportions changed, it would affirm Pavlov's viewpoint, not in terms of the specifics of regulation, but in terms of mechanism—in terms of the ability of the cell to secrete different enzymes at different rates.

The experiment was simple—one stimulant and two enzymes. Were their proportions changed when the rate of secretion was increased by the stimulant? Was secretion "parallel" or was it "nonparallel?" As it turned out, it was "nonparallel." The proportions of the two enzymes were changed in agreement with Pavlov's, not Babkin's prediction!

But my expectation that this would resolve matters, that it would put an end to the old controversy turned out to be dreadfully wrong. As things transpired, these observations were the first shot fired in a long and contentious, indeed, a generational controversy that transcended the dispute between Babkin and Pavlov. In retrospect, their dispute seems quaint.

When I published the results, I found that I had run headlong into what was in some respects the most sophisticated biology of the time. I am referring to what was then a new understanding about the anatomy of the cell and the mechanisms underlying it. The old view was rapidly being rewritten in light of the in-depth examination of cellular anatomy with the high-resolution probe of the electron microscope.

The report of nonparallel secretion produced a far-reaching disagreement between the emerging new biology of the cell, between a whole field and me, an outsider and a novice. As already explained, the dispute concerned the means of protein movement within cells and to the outside. I will have more to say about nonparallel secretion later, but first we need to acquaint ourselves with this new cellular biology, with its beliefs and premises, with what became the standard view of how proteins move. What is thought about these processes today is essentially unchanged from the ideas first articulated some 50 years ago. They are the unyielding science that I confronted as a novice and are the subject of the next section.

*section two*

---

# *On the versimilitude of simulacra*

*chapter five*

---

# The structures
## The cell seen in the
## electron microscope

WHEN I heard the learn'd astronomer;
When the proofs, the figures, were ranged in columns
before me;
When I was shown the charts and the diagrams, to
add,
divide, and measure them;
When I, sitting, heard the astronomer, where he
lectured with much applause in the lecture-room,
How soon, unaccountable, I became tired and sick;
Till rising and gliding out, I wander'd off by myself,
In the mystical moist night-air, and from time to time,
Look'd up in perfect silence at the stars.

**Walt Whitman**
*Leaves of Grass*

In this stanza of Walt Whitman's epic poem *Leaves of Grass*, Whitman makes a comparison between the "charts" and "diagrams" of science and "the mystical moist night-air" and looking "in perfect silence at the stars," thus placing science and nature firmly apart, the former dull and mechanical, the latter, beautiful and transforming. No doubt science is often dry and analytic, and appears divorced from nature's beauty. Indeed, it sometimes seems antithetical to it. Still, the night air and the stars are not only objects of inspiration for poets and other creative artists but also for scientists seeking to uncover nature's secrets. The two are brother and sister on the same voyage of discovery and understanding.

This said, there is one vital difference between the two quests. Critics and historians alike, when on their high polysyllabic horse, say that great art, written, visual, or musical, is found in the verisimilitude of its simulacra. That is to say, it is found in how well its artifacts, its products, its simulacra, reflect some aspect of nature, their verisimilitude. Does a particular work of art inform or enhance our understanding of the world we inhabit and ourselves? Does a written work believably describe feelings, relationships, places, and occasions? Does a portrait painting reveal the person being depicted? Does a landscape or still life painting render the setting faithfully? Does a piece of music evoke sadness, joy, or suffering?

The search for verisimilitude applies to impressionist, surrealist, expressionist, and even abstract artistic manifestations as well as to realistic or representational ones. In any event, it is in such questions and the answers that the creative artist offers that we find art's search for authenticity and through that search, on occasion, greatness.

Unless they are delusional, creative artists never claim that their products *are* what they depict. Their pride is the pride of invention. It is found in the value, meaning, and beauty of certain simulacra. Art, however wonderfully rendered, however enriching, however instructive and insightful is unavoidably fake. It is an artifact. It is not and can never be nature herself. This is not to denigrate the artistic impulse, but merely to say what it is. The greatness of a particular work of art, whether a Caravaggio painting or a Bach cantata, rests on its creation, its invention and production.

Science is exactly the opposite. Great science, true science, eschews artifice. Nature is apprehended directly, its reality exposed. There are no simulacra, and verisimilitude is not enough. Unlike art, science in its truest and most authentic form, in its ideal and deepest incarnation, does not imagine or construct new worlds; it does not produce simulacra. It simply and humbly seeks to uncover what exists.

If the inquiries of the scientist are properly framed and our disposition and abilities allow us to be enlightened, nature lets us in on its secrets. But with this wondrous gift of the human mind comes a critical responsibility. The scientist's duty is to uncover nature's properties as they exist, pure, unfettered, and without preconception. Our wishes, what we hope or expect to find is irrelevant. What we discover is nature's call, and its call alone. Yes, we have hypotheses and theories, guesses about what will be found, but they are just that and must not intrude on our observations. There nature's properties must hold sway.

This distinction between art and science seems unmistakable. A bright line marks it. Art is based on simulacra, while science prohibits them. In these dissimilarities lies the tale we are about to tell. In some respects, a dominant strain of modern biology is like art. It produces simulacra of nature, not pure descriptions of her properties. Unfortunately, in this creative process, humility before nature is often replaced by hubris, as we humans impose what we think should be on what is truly known. The press of belief misshapes and counterfeits observations made on natural systems.

This is problem enough for science, but the difficulty is often enormously magnified by the claim that such simulacra, such forgeries, are in fact true to life and that they reflect nature as it really exists. Disfigured art is misrepresented as science. We can call it an *artful* science. Characterized by a tangle of observation and interpretation, the two often seem to be one in the same. Whether due to ignorance or dishonesty, a glossy patina of interpretation is applied to observations in a way that makes the two seem identical. Interpretation becomes a part of, becomes indistinguishable from observation. They are knotted, and the interpretation of what is observed is believed, however misleadingly, to be what is truly observed. Such inventions can be beguiling, seeming to reveal nature herself.

When this happens, it can be quite a challenge to disentangle observation from interpretation, especially when the deception is intentional, when there is an attempt to conceal and confuse. When distinctions are calculatedly hidden, it is particularly difficult to distinguish fact from fiction, to distinguish what is known from what is merely surmised. The way data are presented, as well as the imprecise and ambiguous use of language, are tools of deception intended to mislead about the actual facts of the matter. When confronted by such research, we must be on guard, especially wary, on the lookout for intellectual ellipses. We must not only be chary about what is said, but also of what is not said but is nonetheless implied.

This elision between observation and interpretation is a central feature of the research we are about to discuss. In telling you about it and my attempts at disambiguation, we start with what at first glance – but only at first glance – is the easiest case, in particular, the structure of the microscopic world, the world of cellular anatomy. What does a particular cellular structure look like? What is its shape and form, what are its contours, lines, and curves?

This seems simple enough. Just describe what you see honestly and accurately. Unfortunately, unlike the world we capture directly with our eyes, we can only apprehend many of the structures in the microscopic world of the cell by adding a large dollop of conjecture and presumption, of prohibited art and artifice to the uncertainties of sight. We usually do not see the object pure and unadulterated. We see only a simulacrum of it. This is often unavoidable, but unavoidable or not, it needs to be taken into account when trying to understand what we are looking at. The problem arises when this limiting circumstance is disregarded in an attempt to substitute one's belief for what is actually being observed.

Our story begins just after World War II, when biologists were given access to a remarkable machine: the electron microscope. It had been developed as a tool for physical research, but had been eclipsed by far more powerful devices, such as the atom-smashing cyclotron, that were far better suited to physicists' desire to examine the intimate nature of matter. As a result, the microscopes sat idle for many years, unused until biologists got hold of them.

Among the electron microscope's shortcomings for research in particle physics was the size of the objects it could see. What it could resolve was insufficient for probing the depths of atomic structure. But what was a shortcoming for physics was just what the doctor ordered for biology. With the short wavelength radiation of the electron beam as the light source, biologists could look at the contents of biological cells at previously unimagined magnification. Even if the subatomic resolution sought by physics was beyond its capability, for biology it improved resolution over the standard light microscope, the sole tool for looking inside cells since Leeuwenhoek had invented it centuries earlier, by an order of magnitude or more.

At the time, a common belief among biologists about the anatomy of cells was that with the exception of the nucleus and certain large structures in protozoa it had none. The cell was an amorphous or unstructured mixture of substances suspended or dissolved in water surrounded by a membrane. This was all that remained of the 19th-century belief that cells contained a living material, the slithering protoplasm. Growing knowledge of the physical state and chemistry of cells during the 20th century had left life's fundamental unit without its protoplasm, bereft, lifeless and characterless. The "living" cell had become the insipid "biological" cell, not much more than a bunch of chemicals in water enclosed by a membrane.

The electron microscope ended this limp, minimalist view of cellular anatomy and cellular life more broadly. Not only was there a gross anatomy of limbs, stomach and so forth that could be seen with the naked eye and a microscopic anatomy with structures large enough to be seen in the light microscope, a new, remarkably rich, complex, and previously unknown or only guessed at world of very tiny structures, a world of ultrastructure, was discovered in cells, thanks to the electron microscope and those who developed it for use in biology.

Though describing these newly discovered structures fully and accurately was the primary goal of the new electron microscopists, their efforts did not stop with mere description, with descriptive anatomy. Quite naturally, they wondered what they had discovered did, and how it did it. What jobs did these structures perform and how? With such thoughts in mind, the new microscopists reintroduced life into the forsaken cell. In contemplating these tiny structures they resurrected the living cell, at least in the mind's eye. Though the material nature of the new living cell was far from the protoplasm of earlier times, the two were kith and kin, vibrant and alive.

The cell was imagined to be seething with life on a miniature scale, not as moving, flowing protoplasm, but as a diminutive Cartesian machine, an effervescent mechanical concoction that would gladden Rube

Goldberg's heart. With feverish activity, even turmoil, and with kinetic urgency, the machine and its microscopic parts moved to and fro carrying out their appointed tasks.

One of these tasks, indeed a central task of the newly discovered apparatus of life, was that of a postman. It moved molecules. Most significantly, it transported the proteins that run life to the correct address within the cell and for secreted proteins to the outside. It was not enough to know the code for proteins written in DNA, or to understand the mechanisms responsible for their manufacture, or even to know what they did, to be useful proteins had to be moved from where they were made to their various and sundry sites of action.

This was not merely important; it was imperative, essential. Without it, the cell would lack organization and without organization, without the polarities of function, without different events taking place at different locations, without topological order, the cell would be little more than a chaotic, lifeless jumble of chemicals. Knowing how proteins were moved from place to place was fundamental to understanding the life of the cell.

The nature of this transporting machine was first imagined over half a century ago in studies using the electron microscope to examine the acinar cells of the mammalian pancreas. The acinar cells are the cells in which the digestive enzymes are produced and stored before being released into the intestinal tract to digest food. They and the duct system they feed account for close to 99% of the gland's mass and form the *exocrine* pancreas. Far better known to most people today, and what the term pancreas often brings to mind, is the remaining 1%–2% of the gland's mass, the *endocrine* pancreas, that secretes the protein hormone insulin into blood.

But as important as insulin is in regulating blood sugar and glucose metabolism, the secretion of digestive enzymes by the exocrine pancreas is of far greater consequence. In its absence, food cannot be broken down. It cannot be digested. And of course without digestion, there can be no assimilation. And without digestion and assimilation, we cannot use what we eat. Quite simply, without digestive enzymes, animals from the tiniest invertebrates to the gigantic whale, from the least to the most complex, would not only not have survived, they would not have come into being in the first place. The evolution of the animal world depends in its entirety on digestion and its enzymes.

Though the structures found in the acinar cell are found in most non-bacterial cells, some are particularly abundant in it. This is often thought to reflect the fact that it produces and secretes large quantities of the aforementioned digestive enzymes. Many of these structures were thought to be part of a complex machine designed to move proteins to their respective destinations within the cell, as well as to the outside. It serves to bring them to one membrane or another, to one organelle or another, or to one extracellular space or another. Similar mechanisms were also imagined for the movement and release of other cell products, such as neurotransmitters, as well as for their uptake from the outside.

In any event, the imagined mechanisms gave rise to the premiere archetype of the new electron microscopy-based view of the biological cell as mechanochemical machine. As it applies to the secretion of proteins, the machine has three major components (Figures 5.1 and 5.2):

1. The endoplasmic reticulum (ER)—orderly stacks of flattened, elongated membrane enclosed sacs that occupy about 25% of cell volume in the acinar cell. The external surfaces of its membranes are festooned with ribosomes, tiny particles composed of RNA and some two-dozen proteins that are responsible for protein synthesis.
2. The Golgi apparatus—also elongated membrane-enclosed sacs, but less prominent. It occupies roughly 10% of cell volume in the acinar cell. Golgi sacs are more irregular in appearance than those of the ER and lack ribosomal adornment. The Golgi apparatus is comprised of four separate structures or parts, each of which is understood to play

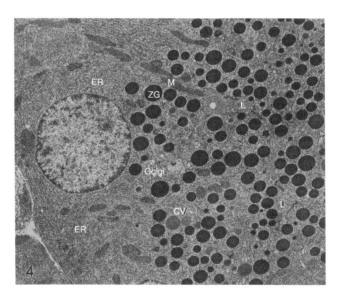

*Figure 5.1* Four acinar cells from a mammalian pancreas. ZG, zymogen granules; ER, endoplasmic reticulum; M, mitochondria; Golgi, Golgi apparatus; L, duct lumen. (Modified from Ermak, T.H. and S.S. Rothman. Zymogen granules decrease in size in response to feeding. *Cell Tiss. Res.*, 214:51–66, 1981.)

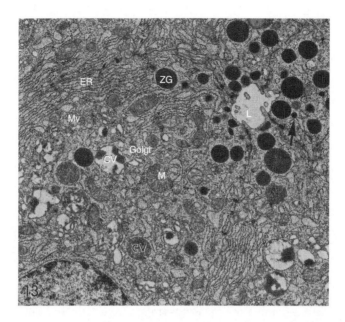

*Figure 5.2* Acinar cell detail. ZG, zymogen granules; ER, endoplasmic reticulum; M, mitochondria; Golgi, Golgi apparatus; CV, condensing granule; Mv, microvesicles (below label); L, duct lumen. To the left of the zymogen granule labeled ZG is a swollen end of an ER sac. (Modified from Ermak, T.H. and S.S. Rothman. Zymogen granules decrease in size in response to feeding. *Cell Tiss. Res.*, 214:51–66, 1981.)

a unique, though often unclear role in the sorting and modification of proteins after their synthesis.

3. Secretion granules—membrane-bound spheres of variable number ranging in size from about 0.2–2.0 μm in diameter that store the substances to be secreted. In the pancreas, they average about 1 μm and fill as much as 40% of cell volume in quiescent or developing glands and as little as 1%–2% after periods of active secretion. Secretion granules were first discovered in the acinar cell in the 19th century (see chapter seven), and were called "zymogen granules," because they were thought to contain zymogens, inactive precursors of digestive enzymes.

In employing these structures, the machine has six features:

1. The proteins to be moved are *contained* within them.
2. They travel *through* them.
3. They are *transferred* from one to the other.

4. This transfer takes place in a *specified order*, from ribosomes attached to the ER where they are made to the internal spaces of the ER, from there in sequence through the various Golgi stacks, and finally from the Golgi to the secretion granule.
5. These structures, including the various Golgi stacks, are *separate* from each other and hence movement between them requires a means of fording the discontinuities.
6. This pathway is *obligatory*. Every last molecule of each and every noncytoplasmic protein follows it and critically is not found elsewhere in the cell except at its ultimate destination.

When I first read the literature on this subject in the late 1960s, I assumed that such a detailed and complex proposal would be based on a substantial body of evidence. What I found was shocking. It turned out that there was precious little evidence, no less of a convincing nature, for the scheme. The whole complex apparatus was little more than an idea, a work of scientific art: it was a scientific simulacrum. If it was not made of whole cloth, it was not much more.

Two things were known. Ribosomes made proteins and secretion granules stored them. Though this was all well and good, it was what took place between these two locations, how new protein got from one to the other that was the task of the proposed machine. Even though the model was relatively clear about how this happened, what actually took place was mostly speculation. To me the whole complex business seemed little more than a figment of fertile imaginations.

Even the anatomical or structural basis for the proposed machine was unclear. Of its three structures, only that of the secretion granule was known with any certainty. And this was not saying much, since the granules were spheres that were clearly separate from each other. Though the many sacs of the ER were thought to form a single contiguous structure, and even though bridges were seen connecting some, the frequency of these links and their ubiquity was unknown.

As for the Golgi apparatus, the notion that it was composed of four separate stacks was little more than guesswork. It might just as well have been a single object as ten objects. One could not even conclude with any certainty that the ER and Golgi were *not* connected. Finally, secretion granules were thought of as outcroppings of the Golgi, and in some characterizations, filled as they formed. But this too was little more than guesswork. It was based on the belief that lucent granules called condensing vacuoles, often seen close to the Golgi stacks, were partially filled zymogen granules (they are not, see Reference 96).

And then there was the topography. It was believed that in moving from ribosomes to secretion granules, the new protein moved from one end of the cell to the other. That is to say, the ER and Golgi lined the path from ribosome to granule. With its attached ribosomes, the ER was located at the blood-facing or *basal* end of the cell stretching outwards towards its center, while secretion granules were bunched together at the other end, at its duct-facing or *apical* surface, where secretion took place. The Golgi apparatus conveniently sat between the two.

Belief in this anatomical arrangement was the basis for the proposed order of events. Though one can find many drawings in textbooks and the like showing this path, and one can compile electron micrographic images showing it, granules were also found at the cell's basal end, and in some circumstances in large numbers, and the ER often extended all the way to the cell's duct-facing surface. And though the Golgi apparatus was usually located in and around the center of the cell, near the nucleus, its scope and boundaries were unclear.

Still, being mindful of these caveats, the arrangement was defensible. It was more or less true. But "more or less true" is not the same as saying that I have five fingers on each hand located in sequence from one side to the other, from thumb to pinky finger. The geometry of these structures, both in their own right and in relation to each other, was not merely uncertain; it was unknown. It had not been established, no less in quantitative terms as needed, and this pretty much remains the case today.

The fact is that what the ER and Golgi look like whole and intact, no less how they are juxtaposed to each other is still a matter for speculation some 50 years after the proposal. As with so much research in this field, the model and unreserved belief in its correctness, buried the uncertainties and speculations. They were submerged in the simulacrum of the model and its attendant guesswork. As a result, and this is critical, what was known and what was imagined became indistinguishable. What was imagined morphed (an apt term) into what was true. Speculation took on an unearned air of certainty.

Finally, as said, the path followed by the proteins was understood to be obligatory. It applied to each and every protein molecule moved by this mechanism. By the same token, passage through compartments *not* specified in the model, the most important of which was the cytoplasm, was *prohibited*. Thus, however insubstantial and unconvincing the evidence, the model not only described a path of movement; it proscribed one.

Most amazing, it could not be said with any certainty that *any* of the proteins were in point of fact in the ER or Golgi. The anatomical evidence for their presence was little more than the occasional (rare) appearance of an undefined dense material in the ER that looked something like the dense material in secretion granules. Against this, the interior of both the ER and Golgi were commonly *less* dense than the surrounding cytoplasm.

Nonetheless, whatever the lack of evidence, belief in the machine was widespread. You could even say that by the time I started to look into the subject it was paramount. Conviction trumped evidence at every turn. Uncertainties were washed away in an ocean of supposition. Observations consistent with the model were welcomed, accepted uncritically, while those that were not were ignored, dismissed, or explained away.

Notwithstanding these deficiencies, those who had interpreted the images were experts, and for most observers, this seemed reason enough to accept their model as being true. I wanted to be one of the believers, but whether constitutionally or intellectually, or both, I found myself unable to join a club based on little more than the authority of certain experts. My refusal was fundamental. In science, an authority's claim should never be taken at face value, as sufficient justification in and of itself for its acceptance. At first, I was not alone in this. Some experts were skeptical, but their concerns were voiced with trepidation and usually in muted tones. However quietly, they averred that the methodological limitations of electron microscopy did not allow for such a grand conjecture, no less a determination of a physiological mechanism.

Most troubling was the fact that the model's inventors were not critics of their own ideas. Instead, they had a vested interest in them. They saw their job as trying to convince others that they were right; that their model described the way things actually happened. As for the images, despite the claim to objectivity, it was no surprise that in their papers and presentations what they chose to show and what they decided not to show reflected their bias.

If this was not enough, their model required that the anatomy that they had envisioned be true and accurate. It was not reassuring that this simply could not be said. It seemed to me that what was being proposed was analogous to trying to establish the existence of the blood circulation from the anatomy of fragments of blood vessels seen in histologic images, without knowing how arteries, capillaries, and veins were connected, or even if they were connected, no less what they contained. The model had been built on an enormous extrapolation from selected details to a grand conception not only of cellular structure, but also of function. As we shall see, as research progressed, belief bettered evidence at every turn.

Not only that, at times in some strange and incomprehensible fashion, the lack of substantive evidence combined with belief in the model's essential correctness to bolster that belief. However peculiar it may seem, on occasion the absence of evidence served *as evidence* for the model. In any event, most of the beliefs that first saw the light of day 50 years ago still endure, and just as surely remain unsubstantiated. This said, neither the passage of time nor the continued absence of important elements of proof, have weakened widespread belief in the ultimate correctness of the model. Indeed, doubtless the idea and theory are more strongly held today than ever. Beyond that, to the community of expert scholars, not to mention the broader community of biologists and laypersons, what they proposed is no longer a model, or even a theory, but an established fact of nature. Fealty to it, as well as to its strictures, has only been strengthened by the passage of time. What was imagined became what is known, and this transformation took place without the interpolation of necessary proof.[6a]

## chapter six

# Microvesicles
## Imagining tiny goings-on

> For the history of our race, and each individual's experience, are sown thick with evidence that a truth is not hard to kill, and that a lie well told is immortal.
>
> **Mark Twain**
> *in an essay Advice to Youth, 1882*

I have just described *what* was proposed to take place, but not the *means* by which it was to be achieved. What mechanisms were thought to account for the movement of proteins from the endoplasmic reticulum (ER) to and through the various downstream compartments to the secretion granule?

It turns out that what was proposed was little more than a series of unsubstantiated speculations based on two key observations. First, the ends of the elongated and flattened sacs of ER lacked attached ribosomes and second, on occasion, they appeared slightly bulbous (see Figure 5.2). This gave rise to the idea that tiny vesicles, called "microvesicles," budded from the ER membrane at this location. This led to a phalanx of some 20 connected speculations.

In particular, if, (1) budding takes place at the ER, then it follows that (2) the vesicles that are formed contain the proteins specified for transport and (3) not others, and accordingly, (4) these proteins must have entered the ER from the ribosomes to begin with. Having budded, the vesicles then (5) carry their contents to the Golgi apparatus, where they are (6) deposited. Then (7) new vesicles are formed from the first Golgi stack, and (8) contain the same proteins, in the same amounts (9). The newly formed vesicles then move, (10) to the next Golgi site. Similar events of deposition, formation, and movement take place at each of the four Golgi stacks in a specified order, 11–22, with the substances ending up in a secretion granule. If we add the mechanisms required to produce these events and regulate their rates, balancing one against the other and against the rates of synthesis and secretion for each individual protein as would be required, the list of speculations becomes very large indeed.

To those not familiar with these ideas, this may seem quite bizarre—so many conjectures from so little evidence. Though the number is remarkable, what was truly astounding was not how many there were, but that they were not seen as speculations at all, but as descriptions of actual events. Like great storytellers who can imagine intricate worlds from a single thought or observation, the electron microscopists had constructed a complex world, along with its dynamics, from two static observations.

However, this was not literature and they were not storytellers, and therein lays the problem. Because they were scientists, the world they conjured came to be seen as real, as substantive. Speculation upon speculation, conjecture upon conjecture, and supposition upon supposition was transmogrified by acts of intellectual prestidigitation into scientific knowledge. The flying broomsticks of Quidditch had materialized!

Yet the fact was that unadorned, the absence of ribosomes at the ends of ER sacs was evidence of nothing more than their absence, and likewise, the occasional swelling was evidence only of swelling. Whatever was believed and however fervently, what, if anything these observations truly signified was unknown. In fact, they might have been anatomical variants of no physiological significance whatsoever. Perhaps the absence of ribosomes was due to the acuteness of the bend at the end of the flattened sac, or maybe their absence and the swelling were artifacts of sample preparation (more about this in Chapter eight). Who could say? What could *not* be said was that such images were proof of budding; no less proof that any of the other proposed events took place.

Even basic anatomical evidence for the idea was absent. For instance, one might reasonably ask, how often were the swellings seen? Did their frequency vary with state? Were there more during periods of active synthesis, fewer when synthesis was inhibited, more during active secretion, fewer during periods of quiescence? Were outcroppings of various sizes seen, suggesting a growing bud? Was the neck of the vesicle seen as it was being formed, and did one see fully formed vesicles sealed off, but still attached to the membrane? Apparently, such things were not seen, or at least were not reported then or in the some 50 years since, no less in quantitative terms.

Nevertheless another anatomical observation gave the microscopists confidence that they were on the right track. Very small, roughly circular membrane-bound profiles of about the same size as the imagined microvesicles, roughly 40 nm in diameter, were actually seen in the cytoplasm. Maybe one could not see the vesicles being formed, but here they were fully formed, sitting pleased as punch in the cytoplasm.

But as with the observations at the ER, these small circular profiles were proof of nothing more than their existence. Who knew what they were or were up to? Indeed, if you wanted to get picky, did you really know that they were vesicles? Perhaps they were balled up fragments of shattered ER or Golgi membrane produced during the preparation of the sample for viewing. Nor were they concentrated near the ER as one might expect. Instead, they seemed to be bunched close to the Golgi apparatus. This was taken as *evidence* that they were either en route to the Golgi or moving their cargo between Golgi stacks. But their proximity to the Golgi might indicate something else. For instance, perhaps they were not vesicles at all, but profiles of circular cross sections cut through elongated Golgi sacs oriented perpendicular to the plane of the section. As with the images of the end of ER sacs, there was no anatomical evidence to help evaluate what they were. Did their number or location change with functional state, with different levels of secretion or synthesis? Whatever they were and what they did (if anything) simply could not be determined from their mere presence.

But none of this mattered. The open questions and lack of evidence were seen as pointless quibbling. Belief that these small, roughly circular profiles were what they were said to be was firm. Like so many tiny buses or jitneys, they carried their assigned passengers from place to place, forming from and fusing with specified membranes (and not others). Proteins that were not stipulated were excluded from the vesicles, while those that were chosen were transferred en route at each stop along the way from bus to bus, from jitney to jitney to their final destination with quantitative precision. But however zealously this was believed, the system with all its bells and whistles was not known. It was only made up, imagined.

When I first encountered this all those years ago, I found it mind-numbing. Not only had an elaborate system been imagined, it had been certified as known based on wisps of evidence. Even at this early time, belief was deep-seated and seemed unshakable. Still, if one could avoid blinking one's eyes, despite the widespread and strongly held belief and regardless of how modish and inventive the notion, it was all no more than an extravagant guess.

It was indeed as real as a Quidditch broomstick. Vesicles were imagined to be budding and fusing, moving to and fro furiously in the absence of substantive evidence that any of this actually happened. Most important of all, observations of the imagined events *in whole functioning cells* was completely absent.

But even if we put all this aside, the concatenation of speculations did not make much sense. To fill one average-sized secretion granule in this way would require hundreds of thousands of tiny vesicles forming, budding, and fusing at each stop along the path, and just as many moving between them. Cumulatively millions of events would be required just to fill one granule. If we add the energy required to do all this to the enormous cost of manufacturing proteins in the first place, the whole scheme seemed, at least to me, highly unlikely, a phantasm, preposterous, and, hyperbole aside, seriously inefficient.

But the difficulty did not end there. The elaborate system that was envisioned was to move proteins an extraordinarily tiny distance. In animal cells, the average distance from the site of synthesis of the protein to its exit from the cell is rarely more than 10–20 µm, a hundredth of a millimeter, a fraction of the thickness of a single strand of human hair. Critically, none of this was proposed because nature demanded it. On the contrary, it was no more than an appealing human invention.

In this regard, *nature offered another, far simpler way to achieve the same end*. Movement could occur by *diffusion*, with the kinetic energy of the molecule *itself* bearing responsibility for its movement. No vesicles, no transportation system, no complex mechanisms, no great energy costs need be imagined. Not only that, but in the natural order of things, it was the first and favored choice for the movement of any and all molecules over microscopic distances.

Why would evolution choose such enormous complexity over the undemanding simplicity of diffusion, and for what purpose? I wondered why the theory's proponents had ignored this simpler alternative when the central rule of science is to favor the simple over the complex at all turns. Scientists are only supposed to be dragged, kicking and screaming, into more complex accounts of nature by the overwhelming weight of evidence, not as seemed to be the case here by the allure or pull of what was thought to be an attractive idea.

The idea's complexity had not earned acceptance as a result of experimental or theoretical test. As said, it was not much more than a product of the human imagination. But more importantly, devotion to the idea was detrimental. It led to weak inductive inferences being used to validate ideas rather than strong deductive tests. "This is what I believe occurs," took the place of "this is what the natural world demands." Human invention, not to mention prejudice, substituted for nature's directive. Alternative points of view were simply ignored.

At first, the model did not have a name. Perhaps its inventors did not think one was needed. After all, they were describing *nature*. However, over the years, a name was attached. For obvious reasons, the conceptual framework I have outlined was designated the "vesicle model" or "vesicle theory." Nonetheless, whatever it was called, in the minds of many model and nature became fused. They were one and the same. As we shall see, this conflation between the two led to an odd, but critical reversal. Observations were thought to support the vesicle model not because nature required it, but because the *model* did. There will be much more about this strange situation later.

In any event, there were two powerful reasons for this state of affairs. I'll discuss the second later. As for the first, scientists usually come to understand nature within the context or confines of their discipline. As students, they acquire the intellectual and technical skills needed to apply its precepts. Whether intended or not, this leads to a kind of indoctrination. The discipline constrains and circumscribes not only *how* its practitioners examine nature but also *what* they examine. This in turn shapes the models they invent.

The investigators here were in great part microscopists, and of course, what microscopists do is look at small things and try to imagine how they function. By its nature, by its tradition, the discipline's ideas are the consequence of what is *seen*. This visual apprehension of nature not only serves as the basis for the experimental work of its practitioners but also informs their thoughts and the models they generate.

As a result, events that cannot be seen in a microscope are often ignored, or treated as if they did not exist, because they lay outside the field of study and expertise of the microscopists and their chosen tools. To imagine proteins moving through the cell by means of diffusion *independent of the structures the microscopists had discovered* was unthinkable. It would not only leave them without structures to use to account for movement, but with those they had so assiduously uncovered lacking known purposes.

To seek the functions of biological cells without reference to their structures made little sense to them. Their task was to figure out the purposes of the objects they had discovered, not to study events independent of them. For the current case, their mission was not to study how proteins are moved *by whatever means it took place*, but to connect structure to function.

Though myopic, this was perfectly understandable. By their own choice, they had limited themselves to understanding the natural world in terms of what could be grasped with a particular tool, the microscope. Their choice was neither nefarious nor useless. To the contrary, it was time-honored. It was what microscopists had successfully done from the time the device was first invented in the 1600s. Their job, the realm of their discipline, was in great part to identify tiny objects *and* guess what they did, linking structure to function by conjecture, by weak inductive inference. As said, this is what I see, and this is what I believe it means. And so, purposes and mechanisms *not* based on anatomical structures were beyond their purview, and they stuck resolutely to their mandate to what they saw.

# chapter seven

# Secretion

## Granules popping out of cells

> So now the Emperor walked under his high canopy
> in the midst of the procession, through the streets
> of his capital; and all the people standing by, and
> those at the windows, cried out, "Oh! How beautiful
> are our Emperor's new clothes! What a magnificent
> train there is to the mantle; and how gracefully the
> scarf hangs!" In short, no one would allow that he
> could not see these much-admired clothes; because,
> in doing so, he would have declared himself either
> a simpleton or unfit for his office. Certainly, none of
> the Emperor's various suits, had ever made so great
> an impression, as these invisible ones.
>
> "But the Emperor has nothing at all on!" said a
> little child.
>
> "Listen to the voice of innocence!" exclaimed
> his father; and what the child had said was whis-
> pered from one to another.
>
> "But he has nothing at all on!" at last cried out
> all the people. The Emperor was vexed, for he knew
> that the people were right; but he thought the pro-
> cession must go on now! And the lords of the bed-
> chamber took greater pains than ever, to appear
> holding up a train, although, in reality, there was
> no train to hold.
>
> **Hans Christian Anderson**
> *The Emperor's New Clothes, 1837*

A mindset similar to that for the microvesicles was at work for the
mechanism of secretion itself. In the mid-19th century, the great German
microscopic anatomist Rudolf Heidenhain noticed that under certain circumstances when pancreatic secretion was stimulated, the area of the
acinar cell occupied by zymogen granules was markedly decreased.[7]
Sensibly, he thought that this had something to do with secretion.

He offered two explanations for the decrease. Either the granules had left the cell or they had become smaller. Though he did not count granule number or measure their size, he concluded that the former was true. Based in great part on the imposing authority of his opinion, this point of view has governed thinking to this day, despite contemporaneous reports that granules shrink.[8]

In the 1950s, there was a change in the way that the loss of granules was imagined. They were no longer pictured, as Heidenhain did, popping out of cells. Now, their disappearance was attributed to a special fusion between granule and cell membrane that led to a hole being produced in both membranes at the site of fusion that served as a channel for granule contents to exit the cell. As the granule emptied, its membrane would flatten out and become part of, indistinguishable from the cell membrane. The granule would vanish as if it had burst out of the cell. This came to be known as *exocytosis*.

A single anatomical observation was the central evidence for this idea. On occasion, a portion of the duct-facing membrane of the acinar cell was shaped something like the Greek letter omega, $\Omega$. This, unsurprisingly, was called the omega figure, and it was thought to be what the cell surface should look like after the fusion of granule and cell membrane. The narrow neck of the omega figure was the hole produced by fusion, and its circular form was the granule itself. In the minds of many, and in many research papers, its presence was seen jointly both as proof of exocytosis and proof that this was the mechanism of secretion. Before long, just about any indentation in the membrane surface of a cell, however broad or shallow, was called an omega figure and seen as proof of exocytosis and the mechanism of secretion.

But as with the end of endoplasmic reticulum (ER) sacs, the omega figure, however suggestive, was evidence of nothing beyond its own existence. It was not proof that exocytosis occurred, no less that it accounted for secretion. It too might be a functionally meaningless variation in the surface topography of the membrane or an artifact of sample preparation. In this light, the omega figure was *not* a common feature of the acinar cell. In fact, it was so uncommon even in actively secreting cells that its discovery required a dedicated search. After one of my first talks about protein secretion at a symposium in the late 1960s, I was asked a pregnant question. "Was there more than one?" In other words, "How many omega figures have you actually seen?"

While I was aware of some examples in the literature, they were few and far between, certainly less than ten. And so, I weakly responded to her question, "I have seen a few." But her point had been made. The duct-facing surface of the acinar cell where the secretion of digestive enzymes into the intestines took place was not littered with omega figures as one might expect, especially in actively secreting glands. They were rare.

The questioner was expressing skepticism about exocytosis and the omega figure publicly, and that was unusual. Whatever one thought in private, what one said in public was that, whatever its frequency, it was sufficient to the task. Though the omega figure was not common, people had come to believe the opposite; that they were common, at least common enough to account for secretion. This was emblematic of the new world of cell biology in which the lack of evidence was no barrier to belief.

At another symposium, a few years later, the creator of the vesicle model and soon-to-be Nobel Laureate gave a talk similar to mine about the mechanism of secretion by the acinar cell of the pancreas. Slides of static electron micrographic images slid by seamlessly, creating the feeling that we were watching a movie, with proteins moving through the various compartments of the cell just as the vesicle model said. The last slide in the sequence was intended to show the final step in the process— secretion. I anticipated seeing the iconic omega figure that he would surely say signified exocytosis. Instead, there was an image of granule and cell membrane fused together. As it flashed onto the screen, there was an audible gasp from the audience. It was not a negative reaction, indicating disappointment due to the absence of the omega figure, but an excited one. It was an affirmation of the compelling new evidence for exocytosis—fused membranes—that they were seeing for the first time.

The idea was that if the omega figure was what should be seen after the formation of the hole in the membranes, then just *prior* to its formation we should see the fused membranes, and here they were as plain as day. Having been chastened about the frequency of anatomical observation in the past, when the time came for questions, I asked the speaker how frequently fused membranes were seen in the acinar cell and under what circumstances?

His answer shocked me. It was not about frequency. Without hesitancy or explanation, the future Laureate said that fused membranes had never been seen in the acinar cell! Well then, what were we looking at? Wasn't the talk about what happened in that cell? He said nonchalantly, as if it were a matter of no importance, that unlike the images in the other slides, this picture was from a liver cell, not the pancreas. He assured me that fused membranes would soon be found in the acinar cell. It was just a matter of time. I don't think they ever were. Indeed, fused membranes seemed far rarer than the omega figure. They were, it seemed, as common as happening upon the skull of an ancient human ancestor while out for an after-dinner stroll in Berkeley. The fact was that, despite many granules sitting cheek by jowl with the apical membrane of the acinar cell, *none* was fused with it, even during very active secretion. Nonetheless, we were asked to believe that this single, apparently singular image was somehow representative.

By this time, I had become a full-fledged skeptic. I thought that the image of fused membranes was no more evidence of exocytosis than the omega figure. But there was something else about his final slide that I found particularly disturbing. Why the deception? Why, when every other image had been from the acinar cell, at this critical juncture were we shown a picture from a completely different cell *without note*, with the apparent intention to deceive?

The reason for the sleight of hand was obvious. Quite simply, if fused membranes were powerful evidence for exocytosis, then their absence argued just as strongly that it did not occur. The image was proof for the exocytosis hypothesis, in the liver if not the acinar cell, and was chosen to make this point. This made the doubter in me wonder: Weren't all the slides we had been shown chosen to make a point? In fact, hadn't everything we had seen been selected to make a point favorable to the vesicle theory?

The answer was "yes," and this was undeniable. Where, I asked myself, were the pictures that were not so salubrious, that pointed in another direction? Were we to believe that none existed? The speaker had presented the audience with what he believed to be the *best* available images, but the "best images" were those he had handpicked that "best" fit his theory. The whole business seemed concocted. We had, I thought, been treated to a system of belief supported by strong convictions, not by convincing evidence.

But I seemed to be alone in these feelings. Those in attendance seemed delighted by the picturesque story and the powerful new evidence of exocytosis. To question the story, to point to pimples on the airbrushed face of the theory, was not just to be a curmudgeon, it was to be an apostate.

Like it or not, the virtual absence of proof for exocytosis was no small matter. If it did not occur, or if it could not account for secretion *en toto, then something else did.* And therein lay the problem for the vesicle theory. The only possibility for that *something else* was that proteins left the granule and cell by passing through the interposed membranes and cytoplasm en route. And if this was true, then there was no sequestration of the digestive enzymes in the ER and everything that was said to follow was not to be believed. The whole story would collapse of its own ponderous weight.

I expected that others would share my skepticism and was surprised when they didn't. After all, exocytosis and its proposed role in secretion was just an idea, wasn't it? Just a possibility to consider? Who could say whether or not it was true? But this was no ordinary situation. Despite the absence of evidence, it was widely believed that exocytosis was responsible for protein secretion by the pancreas, and for that matter, for the secretion of protein wherever it occurred. In turn, this belief gave credence to the vesicle transport system with its sequestration and microvesicles, while at the same time, belief in sequestration and microvesicles gave credence to exocytosis. It was all beautifully circular.

## chapter eight

# To see, we must transform
## What are you really looking at?

> A philosopher who has been long attached to a favorite hypothesis, and especially if he has distinguished himself by his ingenuity in discovering or pursuing it, will not, sometimes, be convinced of its falsity by the plainest evidence of fact. Thus both himself, and his followers, are put upon false pursuits, and seem determined to warp the whole course of nature, to suit their manner of conceiving of its operations.
>
> **Joseph Priestley**
> *The History and Present State of Electricity, 1775*

This is not the end of the difficulties for the vesicle model, not nearly. What were the electron microscopists really looking at? Thus far, we have talked of the images as if they were in something approaching their natural state. But this was not so – far from it. They were not even inert and motionless remnants of what existed in the living cell. In fact, they were not even of nature's making. What was being looked at was an artifice. It was the microscopist's, not nature's handiwork.

Studying the cell and its contents in the electron microscope not only required its modification but its complete transformation. Though rarely thought of as such, what was actually being examined was a *cast*, a replica of the natural object. The microscopist's hope was that the cast was true to life, that it was a versimilitudinous simulacrum.

What they had done was remarkable. It was no easy task to create a faithful cast on such a small-size scale. Still in the end all that could be said about their constructions were that they probably looked something like the natural object. This is our best guess, our *hypothesis* for the natural state. There was ample reason for thinking that it was a good guess, a good hypothesis, but whether it was good or bad, and however well rendered, obviously a cast of something can only be a model. It can never be the thing itself.

Let's briefly go through the transformation that took place from natural object to cast. As with traditional light microscopy, the microscopist's first task is to "fix" the sample. This is an attempt to retain the structures' natural forms and relationships and to prevent their movement or distortion during subsequent preparative procedures. Fixation seeks to join the contents of the cell and the object more generally to each other, to, in essence, make them immobile by making them into a single object.

This linkage occurs at the molecular level and, for most "fixatives," is the result of the formation of chemical (covalent) bonds between the cell's various parts. In the light microscope, this is thought to have little to no bearing on what is observed, because the objects are being examined at relatively low magnification, far away from the molecular level where the bonding occurs. Whatever distortions or displacements take place, are simply not seen.

But the situation is very different for electron microscopy. Here the size of what is being examined approaches molecular dimensions, and even minute displacements or distortions, including the fragmentation or coalescence of parts, can completely alter the appearance of the object and the meaning of what is being looked at. Even more problematic, there is no certain way of knowing whether changes in shape and location have occurred and, if so, are of a magnitude and nature that is anatomically and functionally consequential. As we shall see, this problem is particularly acute when we consider the features of the vesicle theory.

What was missing was a frame of reference. Except for certain cases, the electron microscopists had no standard for comparison. They did not know how the things they were examining looked, or what their location or relationships were, from some other source of information. They couldn't compare their casts to *known* structures because they were in the process of discovering them. Many were being seen, or at least seen clearly, for the first time.

In any event, fixation was just the first problem. To examine a tissue in a microscope, it has to be sliced into sections thin enough to be transparent to the incident light. This is true for all transmission microscopy, where light transmitted through the object forms an image, but this requirement is particularly demanding for electron microscopy due to the extremely short wavelength of the electron beam. Its "light" is completely absorbed by samples as thin as single cells, thereby making them opaque. With no light being transmitted through the object, what is inside cannot be seen.

To overcome this obstacle, a remarkable little machine called an ultramicrotome was invented. It cuts sections of tissue so thin, no more than slivers of cells, that they are transparent to the electron beam. But nature extracts a great price for this accomplishment. As a result, many cellular structures are only known in terms of thin slices through an unseen and unknown whole. To determine the anatomy of the whole object from such shavings requires an extrapolation not only from parts to an unknown whole but also an extrapolation from parts of those parts. This leaves the microscopist unable to do more than *imagine* what the natural object, whole and unaltered, looks like. The actual anatomy of the structures and their relationships to each other cannot be directly observed. All we have are hypotheses of how they might look.

And developing credible hypotheses of the whole from these shavings is not an easy task. It is something like trying to determine the structure of an automobile from a slice through say 1/20th of its thickness, and here is the kicker, not knowing what a car is, that it has wheels and a motor, or even that it moves. To say the least, what we would end up with would be little more than guesswork, and without the introduction of some other frame of reference, would remain so in perpetuity.

If this weren't challenge enough, before we can examine our thin section in the electron microscope, we must remove and replace all of its water, some 80% of the cell's contents. This is because the high-energy electron beam vaporizes water. The material we replace it with must be robust enough to withstand the insult of the beam of electrons. In addition, to be sectioned by the knife of the microtome, it must be solid. Trying to cut a squishy object with a knife, no less into microscopically thin slices, would be a fruitless, and an exercise in frustration.

The replacement of water with a solid material must, of course, be made without distorting the structures or changing their locations and relationships. To do this, a process of gradual substitution is usually used. In it, the water is purged, and a plastic, in liquid or monomeric form, takes its place little by little. When the substitution is complete, the plastic is hardened or cured, and the cast is produced.

And if all this is not sufficient, there is one final hurdle. It is difficult and at times impossible to distinguish one object from another in microscopic sections. The reason is that there is little contrast between them. This is because they are all composed of the same chemical elements, centrally, carbon, nitrogen, oxygen, and hydrogen, and thus are of roughly the same density, with the same absorptive properties. And yet, to make useful observations, we must be able to distinguish one from the other.

Doing this is task of "stains." Stains are chemicals that when added to the sample, accentuate one structure or one aspect of it relative to others. The most commonly used stains for electron microscopy are salts of the heavy metal osmium. But as with everything else in this process stains can produce distortions. In fact, it is the stain that we are actually looking at, and it is by definition an artifact, something that we have added to the sample.

Taken together, these circumstances may seem daunting to the unini-
tiated. How can we learn anything of substance about the intimate parts of
cells in the face of such massive alterations? Moreover, since the structures
were unknown to us to begin with, how would we ever know whether we
had been successful in reproducing the natural state?

The way in which all this uncertainty was dealt with is instructive.
With time, the community of electron microscopists came to agree about
what constituted a "good method" of preparation. A consensus was
formed about which methods produced the most faithful rendering of the
natural state. Agreement was reached at meetings, through the editorial
boards of various scientific publications, and by means of the review processes of government agencies.

One critical element in forming this consensus was invariably left
unspoken. It was aesthetic. Geometric order and a beautiful appearance
(in the eyes of the beholder) were understood to be more reflective of the
natural state than a chaotic or ugly appearance. This, even though in
nature beauty and order often underlays a chaotic and ugly appearance,
and conversely geometric order can be artifactual, imposed by a particular treatment.

Nonetheless, as said, despite all the uncertainties, there was often
good and considered reason for thinking that what had been agreed upon
looked something like the natural object. This is an enormous compliment
to the great care that the developers of these methods took in choosing
and applying various procedures. Still, however comforting, it is important to keep in mind that the methods were developed with a strong drive
and even more passionate desire to uncover what was there to be seen.
Finding it was more important than the accuracy of the transformation.
There seemed to be no choice in this. To make progress and develop a
reproducible procedure, a judgment had to be made about which methods
were satisfactory, which came closest to retaining the natural state and
which did not, however uncertain that judgment was.

But even if all these hurdles were cleared, three great uncertainties
remained. First, the form, as well as the location and relationships of
the most intimate details of the objects, *those approaching the molecular
level* were unavoidably ambiguous. Two, at the other end of the spectrum of size, cellular organization in general was uncertain due to the
fact that it was not possible to examine the larger structures and their
relationships to each other, whole and intact. For instance, the structures and arrangements of the endoplasmic reticulum and Golgi bodies were little more than impressions drawn from thin slices through
the objects. Finally, third, without a *quantitative* description of the
cell's anatomy, whatever conclusions were gleaned were necessarily
incomplete and uncertain. It was and remains absent. The dream of an

accurate description of the cell's anatomy remains unrequited. Much has been learned, but a great deal of what is thought is no more than a widely agreed upon guess. Sometimes that guess is well informed, but just as frequently it is contaminated by bias, by preconceived notions about how things should look. We will see how this worked when we consider how the vesicle theory and its proposals were superimposed on observation.

## chapter nine

# Making things whole
## To get beyond speculation

> False facts are highly injurious to the progress of
> science, for they often, endure long; but false views,
> if supported by some evidence, do little harm; and
> when this is done, one path towards error is closed
> and the road to truth is often at the same time
> opened.

**Charles Darwin**
*The Descent of Man, 1871*

Even if we accept the vesicle theory's view of everything, there is one
shortcoming that cannot be avoided, managed, or explained away. All the
evidence in its behalf was inescapably static. *None* of the newly described
structures could be seen doing *anything*. Regardless of how carefully and
faithfully the living cell had been transformed, what was studied was a
lifeless artifact of man's construction.

However wonderfully evocative and beautiful the images, however
vibrant and inventive the thoughts about their meaning, however percipi-
ent the intuition and trustworthy the intention, the Cartesian machine of
the microscopists' imagination, with its frenzied movements, remained
just that, a figment of the imagination. Its vivid dynamism only existed in
the mind's eye. This was indisputable.

Yet—and this is what was so remarkable—a complex cellular machine
*along with its dynamics* was not just imagined; it was widely believed to
exist. For most observers, the reality was obscured by enthusiasm for the
model. Only the scarce few not caught up in its conceptions and conceits
were able to see just how little was truly known. Missing, and of self-
evident and vital importance, was knowledge of what actually took place
in the intact functioning cell, in the living cell!

That this huge deficiency was not a significant impediment to belief
was in great part a reflection of the discipline of microscopy itself. As
explained, examining structure and inferring function was a common
practice among microscopists. Though many of their inferences were
thoughtful and eventually confirmed by other evidence, many were not.

Theirs was primarily a guessing game. However perspicacious, and with a few notable exceptions, all they could offer were hypotheses for what took place.

This left some electron microscopists hesitant to draw functional conclusions from their images. They understood the limits of their methods and were uncomfortable guessing about what they saw did no less how it did it. Elsewhere I have called them "conservatives" because of their reluctance to believe in a world whose existence was mainly aspirational.

I have called those who were willing to speculate, to infer function from structure, "progressives." They understood that, whatever the limitations of their speculative inferences, they were necessary if one hoped to make progress. Just as the caution of the conservatives was justified, so too were the progressives justified in thinking that their speculations were needed. If nothing else, as Darwin suggested in the epigraph heading this chapter, they would help set the path for future more definitive research.

The vesicle theory was the most elaborated and complex of these speculations. As already noted, the great danger was that what was introduced as an idea could absent further evidence or proof quite easily come to be seen as a fact of nature simply as a result of the expressive power of ideas. Some progressive microscopists were aware of this risk and appreciated the need for evidence of what *actually* took place in cells as a bulwark against mere speculation. Among these thoughtful progressives were the vesicle model's inventors.

To their credit, and despite having the utmost confidence in their model, they understood the importance of obtaining dynamic evidence in the living cell, and sought to do just that. But praise for their effort must be tempered by what they actually did. As it turned out, their experiments were *not* designed to test their theory against the properties of secretion as seen in whole, functioning cells. They were calculated to do one thing and one thing only—to bolster or buttress their theory, to *reinforce* its speculations. There was no possibility that their experiments would lead to the theory's rejection.

We will get to why in a moment, but despite this fundamental shortcoming, their experiments came to be highly regarded not only as scientific achievements, but also as an estimable approach and method of research in cellular biology. In fact, it is fair to say that they became the gold standard for the study of the biological cell, as important to the new paradigm as the vesicle model itself. If one hoped to learn about the biology of the cell, the approach they had pioneered was not merely useful; it or something like it was *de rigueur*.

The key difficulty with their experiments was, as said, that they did not actually measure the properties of secretion in the living cell, though they claimed to have done just that. Before we look at the experiments, we need to set the stage.

Their studies were a reflection of a broad shift in the mindset of biologists or perhaps more accurately, a change of heart that took place after World War II. For more than half a century, biochemists had had a great deal of success discovering the chemicals of life and establishing their reactions and interactions by breaking cells open and studying their chemical constituents in isolation, in test tubes, outside of, and separated from the enormous complexities of their cellular source. To do this, tissues and their cells were "homogenized," ground up and blended. This was usually accomplished with shearing forces applied with a rotating blade or rotating pestle (fit tightly into a mortar shaped like a test tube).

The result was a milky suspension of the cell's contents that served as the starting material for chemical purification. By the end of World War II, the success of these "grind and finders," as they were sometimes disparagingly called, had become undeniable. Not only was it possible to learn about the chemistry of the cell by destroying it; this seemed essential. To obtain a rigorous understanding of biological chemicals they had to be held, so to speak, in your hand, isolated from the cell and from other molecules, pure and unadulterated.

The change that took place in the wake of World War II was to extend this approach to the tiny structures uncovered in the electron microscope. They were part of the milky suspension of homogenized cells, and isolating them from the medium and from each other seemed like a natural extension of what had been so successful with biochemicals, and many thought it would be just as fruitful. The hope was to study their properties in isolation, just as so many chemical compounds had been studied.

To some, this hope was completely delusional. How could the physiology of cells be understood by destroying them? That might work for chemicals, but homogenates had no anatomy and no physiology. To get at what was inside, it was necessary to destroy cellular organization, and while you might learn something this way, how the isolated structures worked in the intact cell was not one of them. To expect to gather meaningful information about function from this ungodly mishmash was absurd.

But before long, this point of view not only came under fire, it came to be seen as superannuated, as backward and primitive, old-fashioned, and even unscientific. In any event, for both good and bad reasons, biologists set about separating the tiny structures in the suspension from each other mostly using centrifugal force, sedimenting different objects at different speeds based primarily on their size, weight, and density.

Many believed that just like the cell's chemicals, the functions of these structures—what they did and how they did it—could be, and indeed could *only* be fully and properly understood by studying them in isolation. This would permit the researchers to learn about them in ways that were just not possible in the complex and often, figuratively and literally, opaque whole cell. The traditional approach seemed naïve, unsophisticated, and was thankfully now out of date. Modern science could be far more thorough, more precise, and provide far deeper understanding. By studying the isolated structures, we could determine in intimate detail how they functioned, molecule-by-molecule and chemical property-by-chemical property. There would be more great achievements for reductionist biology to add to its already impressive list.

The utility of this approach had previously been shown for the cell's largest inclusion, the nucleus, and extending it to smaller structures made perfect sense. And there was one initial enormous success. It concerned protein synthesis and the tiny ribosome. Ribosomes are relatively easy to separate from the homogenate, at least in principle. All one has to do is remove larger structures at lower centrifugal forces and then bring the floating ribosomes down at ultrahigh speeds.

Ribosomes isolated in this fashion were, under the proper conditions, able to make proteins. This opened the door to deciphering the process of protein synthesis in intimate detail in a test tube, something that would have been extremely difficult if not impossible to do in the intact cell. Critical to the success of this approach was the very small size of the ribosome. Most survived homogenization *intact*, slipping easily between slicing blades and crushing pestles. Deciphering protein synthesis in a test tube this way was both impressive and very important, but as it turned out, the approach was not general and could not easily be applied to other important cellular structures.

This brings us to the experiments in question, concerning the secretion of digestive enzymes by the pancreatic acinar cell. To assess the dynamics of the secretion process, the inventors of the vesicle theory sought to extend the methods of homogenization and separation to two structures that were orders of magnitude larger than the tiny ribosome, the endoplasmic reticulum (ER) and the Golgi apparatus. The problem was that both were hopelessly fragmented by homogenization. What was collected in the centrifuges were their remains, their remnants—small membrane-bordered or enclosed objects that had been part of the larger structure before it was broken apart.

Residua with ribosomes attached to the membrane were presumed to be fragments of the ER and were named "microsomes." Identifying pieces of the Golgi apparatus was more difficult and uncertain because they lacked such an anatomical identifier. Homogenates contained all sorts of small ribosome-free, membrane-bordered, or enclosed objects that looked much the same. No doubt some were pieces of the Golgi, but others were doubtless fragments of other structures destroyed during homogenization, for instance, the cell membrane, membranes of broken secretion granules, lysosomes, nuclei, pieces of the ER denuded of ribosomes, and if they existed, of course microvesicles. Add that separating fragments of the various Golgi stacks *from each other* seemed even more problematic and you can appreciate the staggering difficulty researchers faced in attempting to understand what happened in the intact cell by examining these isolated scraps.

Even if all this could be sorted out, they would still only have bits and pieces of what they had demolished. If the hope was to examine the role of these structures in the process of secretion by examining them in isolation, it simply could not be done. To study the object, they had to destroy it as well as its critical relationships. This was just as the naysayers said!

However, this seemingly crushing limitation did not dampen the enthusiasm or optimism of the investigators. They thought that, with insight and ingenuity, with wisdom and intellect, and I would add more than a pinch of hubris, they could overcome such difficulties. They would establish the properties and relationships of the natural objects as they functioned in the intact cell from their ruins, from these tiny bits and pieces. They thought that they could put what they had rent asunder back together mentally if not physically. It was as if an archeologist not only hoped to establish the layout of an ancient city and the architecture of its buildings from their ruins, but resurrect the people who inhabited it going about their daily business.

In pathbreaking and as noted, frequently copied experiments, they were thought to have achieved this dazzling goal. They had, it was thought, been able to discern the dynamic processes that moved proteins from place to place in the intact cell from the study of fragments. And yet, despite the claim to the contrary, the bald fact was that not only hadn't they realized their dream; its realization was not possible. All they ended up with was a giant misapprehension.

Critical to the conclusion that they had succeeded were various assumptions. If they were warranted, then, yes, they had actually measured what took place in the cell. If they were not, then not only wouldn't they have achieved their goal, but even worse, their results would be uninterpretable and meaningless.

But before we consider the assumptions that were made, we need to understand what they were *not*. They were not general assumptions about nature or even about biology. Rather, they were specific to the vesicle model, drawn from *its world*, not from nature. This resulted in concluding *a priori*, that is, before they carried out their experiments, that things happened in much the way that their model proposed. Their assumptions, not evidence, not nature provided the ground rules for understanding. They determined what could occur and what could not.

In doing this, they conflated their model with nature. In a serious case of circular reasoning, they assumed that things took place more or less the way their model said, and then concluded from their observations that this was true, that what they had assumed to be true was true. This was truly astonishing. Not only wasn't the model being tested, the truth was thought to be in hand before the experiments were carried out. This unfortunate state of affairs was unavoidable. It was the result of examining the parts of the cell in isolation. As a result, their experiments could only be understood in light of, not independent of, the model's constraints and assumptions. That this was the case was obscured as they tried to make the cell whole by pasting its parts together with their model and its assumptions as the glue.

With assumptions masquerading as facts, the fragmentary and uncertain nature of the data became irrelevant. If the assumptions were appropriate, it all fit together fairly well. But, and of course this is critical, if they were not, if they could not arrogate nature to the demands of their model, the meaning of their observations was not only unclear but more than that, as said, they were uninterpretable, at least rationally. Indeed, if one accepted their assumptions, the truth was known beforehand and their experiments were unnecessary. All they did was confirm what was already known.

Even with this major problem, it is hard to overstate the impact that these experiments had on the study of the cell. The common impression was that great progress had been made. To me, their influence was profoundly *negative*. Research in cell biology had been infected with the horrible idea that it is perfectly acceptable to conflate nature with man-made models of nature.

The idea was insidious. It gave the impression of deep understanding, when all one really had was a hidden confusion between the model and the natural world. Sometimes, the conflation was so great that the two were actually fused, indistinguishable. Fused or not, what was presumed became illuminating fact.

*Interpretation had not been used to inform evidence, the hope of all scientific interpretation; it became evidence.* If you find this confusing, I hope to unravel it for you in the following chapters. Anyway, however illustrious, however fêted, these studies set the field of cell biology on an inexorable path to entrenched belief, to dogma—*with models, not nature, leading the way*. A new biology of the cell had come into being all right, but encumbered by a deeply flawed sense of the relationship between data and interpretation. Conflating what was assumed with what was observed became all too common.[9]

## chapter ten

# The search for dynamics
## The theory examines itself

First, if any opinion is compelled to silence, that opinion may, for aught we can certainly know, be true. To deny this is to assume our own infallibility. Secondly, though the silenced opinion be an error, it may, and very commonly does, contain a portion of truth; and since the general or prevailing opinion on any subject is rarely or never the whole truth, it is only by the collision of adverse opinions that the remainder of the truth has any chance of being supplied. Thirdly, even if the received opinion be not only true, but the whole truth; unless it is suffered to be, and actually is, vigorously and earnestly contested, it will, by most of those who receive it, be held in the manner of a prejudice, with little comprehension or feeling of its rational grounds. And not only this, but, fourthly, the meaning of the doctrine itself will be in danger of being lost, or enfeebled, and deprived of its vital effect on the character and conduct: the dogma becoming a mere formal profession, inefficacious for good, but cumbering the ground, and preventing the growth of any real and heartfelt conviction, from reason or personal experience.

**John Stuart Mill**
*On Liberty, 1859, p. 72*

We are now ready to consider the particulars of the experiments. The key study sought to follow a wave of radioactively labeled protein as it traveled through the cell. According to the vesicle theory, the radioactivity would first crest on ribosomes where the protein is made, then, in order in the endoplasmic reticulum (ER), microvesicles, the Golgi apparatus, and finally secretion granules.

As already pointed out, what was actually being followed was not a wave of radioactivity or anything else *in the living cell*. The measurements were of variations in the amount of labeled protein in particular *subcellular fractions* collected from homogenized tissues that were to serve as *proxies* for the ribosomes, the ER, the Golgi, microvesicles, and zymogen granules in the intact cell.

The experiments began with a radioactive substance being incorporated into new protein, either in the pancreas in situ or in pieces of pancreatic tissue in vitro for a brief period of time. At certain times after this exposure, the tissue was homogenized and specified fractions collected by centrifugation. The rise and fall in the amount of radioactive protein in each fraction was then measured and the fluctuations assumed to mimic events in the intact cell.

As said, others had already determined that proteins were manufactured on ribosomes and stored in secretion granules. So the presence of labeled protein in these structures and their movement from one to the other with time was no surprise. The question was not whether this transfer took place, but what path was followed. This was what the experiments were intended to discover. Did the protein move through the ER? Was it carried in microvesicles to and through the Golgi? And how exactly was it transferred from the Golgi to secretion granules? Did things occur in the way the vesicle theory envisioned?

Because they were *not* following actual events in the living cell, it was necessary to establish that each cell fraction was a proper surrogate for the intact structure it was thought to represent. In other words, did changes in the amount of radioactive protein in each fraction reflect events in the living cell? Ribosomes and secretion granules aside, were they appropriate surrogates? Herein lies the problem. This was never established. Instead, the fractions were simply *presumed* to be satisfactory stand-ins.

This was based on three assumptions. The first and most obvious was that the presence of the labeled protein in fractions associated with the vesicle model reflected their presence in the comparable compartment in the intact cell. Or put the other way round, their presence in these fractions was not an *artifactual inclusion* produced by homogenization. Second, it was assumed that the labeled protein was held *within* the collected objects, and was not merely adsorbed or otherwise attached to their surface, or suspended in the surrounding medium. The third assumption was that the presence of labeled material in cell fractions *not* associated with the vesicle theory was artifactual, due to the redistribution of labeled protein during homogenization. In this case, its presence did *not* reflect what is found in the intact cell.

Given the correctness of these assumptions, one could say and say with great confidence that the labeled protein content of the approved fractions reflected what took place in the intact cell. More than that, it could be said that the measurements validated the vesicle theory. They proved that it was correct, that it provided a true and accurate description of nature's mechanism. The problem was that if the assumptions were justified, then the vesicle theory was not a theory at all, but an established fact of nature *known before the experiments were performed*. If they were justified, the facts were already known and the theory was acknowledged to be true beforehand.

As a result, the experiments were superfluous. However, if the assumptions were not justified, *without* their favor, the whole conceit falls apart and *none* of the conclusions would be justified. The relationship between what was measured and what took place in the cell would not merely be unknown; it would be unknowable. The data would be uninterpretable.

Arguably, the most consequential assumption was the third—the exclusion of the labeled material from compartments not specified in the vesicle model. In this regard, the supernatant fraction of the homogenate, the material that did *not* sediment, was critical. It contained the contents of the cytoplasm, and in their experiments accounted for about 20% of the total radioactive protein in the tissue. Others have reported values as high as 60%. But whatever the number, the assumption was that every last molecule of labeled protein found in the supernatant fraction was an artifact of redistribution produced during homogenization, from broken ER or Golgi sacs, secretion granules, or from duct fluid, *not* from the cytoplasm. If one could examine the cytoplasm in its pure state, it would be totally free of labeled protein.

This assumption by itself made the vesicle theory true *a priori* and the experiments pointless. If one could conclude that, on leaving the ribosome, the new protein molecules *did not* enter the cytoplasm, the only possible alternative (proximate) destination was the cisterns of the ER. They had no other place to go. And if, having entered the cisterns, they could *not* leave on their own accord to enter the cytoplasm, then how else could they move about the cell except by being carried from one membrane-enclosed compartment to another in microvesicles or the like? With exclusion of the cytoplasm, vesicle transport was the only means of moving proteins from place to place. It made the vesicle theory true ipso facto. Otherwise, new proteins would be trapped on the ribosomes immovable, and clearly this was not the case.

And so with a swipe of the assumptive hand, the vesicle theory was left unopposed, proven true by default. The only other possible explanation, cytoplasmic passage, was excluded *a priori*, not on the basis of evidence, but presumptively. The investigators seemed to have appreciated this difficulty and attempted to corroborate their observations in a parallel series of experiments in *intact cells* using a method known as autoradiography. Autoradiography had previously been used to identify the location of radioactive substances in microscopic sections at the level of magnification of light microscopy. Their plan was to apply this method to the far more demanding circumstances of electron microscopy.

In autoradiography, a film emulsion is placed directly over sections of tissue prepared for microscopy. In the current case, they were placed over the ultrathin tissue sections used for electron microscopy. As in the cell fractionation studies, radioactive protein was produced in the living tissue for a brief period of time, except that instead of being homogenized, the tissue was fixed, sectioned, and readied for microscopic viewing.

The idea was that radioactive proteins in the thin tissue sections would emit radioactive particles that would expose the emulsion and produce dark spots or grains on it whenever and wherever they landed. By examining tissue sections at different times after the radioactive protein was manufactured and determining the location of the grains on the exposed film, they could follow changes in its position in the cell. All they had to do was follow the dots.

Did they move from the ribosomes to the cisterns of the ER, to and through the Golgi stacks, and finally to secretion granules? Could the inferences made in the cell fractionation studies be confirmed in intact tissue sections? If they could, then taken together, the two sets of experiments would, it was thought, be powerful proof for the vesicle theory.

The approach seemed utterly straightforward—grains were either here or they were there. But as is often the case in science as in life otherwise, things were not so simple. There were difficult and in some cases limiting interpretive problems. Autoradiography introduced new uncertainties and ambiguities, its own questions of artifact and distortion. To begin with, the location of the grains on the emulsion was *not* its uncomplicated superimposition from the underlying tissue section. When a radioactive particle is emitted it rarely follows a straight-line path perpendicular to the point of emission (superimposition).

Its route is random and expanding. As such, it might land on the emulsion far away (at some considerable lateral distance) from its point of origin in the tissue section. This distance is a function of the energy of the particle—the higher its energy, the further away from the source it can travel before encountering the emulsion. It is also a function of the distance it has to travel to reach the emulsion, the greater the distance, the greater the potential displacement. This in turn is determined by two factors: the thickness of the tissue section and how closely apposed it is to the film.

Thus, depending on the energy of the isotope and how far it has to travel, the particle will intersect the emulsion at some lateral distance from its point of origin. The grain may be found over object X, when it was emitted from object Y next door, down the road, or even around the corner. The potential paths that the emission can follow form a cone whose diameter increases as the particle moves further away from the source. The base of the cone (its greatest diameter) is formed when the emission encounters the emulsion. This is called "the resolution element." The source of the particle in the tissue section may be anywhere within the resolution element.

If the resolution element is smaller than the object in which the grain is observed, then it can be said that the particle originated from it. If it is larger, then it may come from elsewhere, from the cytoplasm or some other nearby structure. In their experiments, location was a major source of uncertainty, and was dealt with, as in their other study, in a fashion designed to bolster, not challenge the vesicle theory.

For instance, the impression was often given that a radioactive emission came from a particular structure when its true origin could not be determined. At times, with the turn of a phrase, this indeterminacy was simultaneously admitted to and denied. Grains were said to come from a certain "area" or "region" of the cell, while at the same time, it was implied that they came from a particular object in that area or region when that was simply not known.

For example, it was common practice to say that grains found over the "area" of the ER came from within its cisternal spaces, not ribosomes, not the external surface of its membranes, not other nearby structures, and most importantly, not the adjacent cytoplasm when these distinctions simply could not be made. Similarly, it was implied that grains found near or over microvesicles were emitted from them, when it was just as likely, or perhaps even more likely, given their small size that they came from the nearby cytoplasm or some other nearby structure.

I remember asking one of the developers of the vesicle theory during a seminar he was giving what he meant by "region" of the ER. The way he responded highlights this point. At first, he agreed with me that he could not say with certainty that the emissions came from the internal spaces of the ER. The resolution of the method simply did not allow for such a conclusion. But before you could blink an eye, he added that he was quite sure that this was where they came from.

There was another interpretative difficulty with autoradiography that was equally problematic and that was all too often ignored. The number of grains on the emulsion varies with the time of exposure, just as with the darkness of an ordinary photograph. Naturally, the longer the film is exposed, the greater the number of emissions, and hence the greater the number of grains on the emulsion. With enough time, there would be multiple emissions from each labeled molecule. In fact, if one waited long enough, there would be so many overlapping grains that counting them would become a difficult, if not impossible task.

On the other hand, if the time of exposure were short enough, grains would only be found over compartments that contain the proteins at a high concentration, where the likelihood of emissions is great. As a result, a small compartment that contains the radioactive protein at a high concentration would be seen in the film or autoradiograph, whereas a large compartment that contained far more labeled protein, but at a lower concentration might appear empty, free of radioactivity.

Once again, the cytoplasm enters the analysis. Due to its large volume, about 50% of cell volume, even a large amount of radioactive protein if present at a relatively low concentration, might be missed. As a result, one might draw the erroneous conclusion that there is no radioactive protein in the cytoplasm, when in fact it was home to a great deal.

At least in theory, there is an ideal time of exposure. One spot would be produced for each source protein, no more, no less. However in practice, the time of exposure was the investigator's choice. It was his or her preference. This made the measurements judgment-laden. Quite naturally, the investigator chose a time of exposure that gave the best result. And also quite naturally, the best result, once again, was the one that best fit his or her hypothesis.

Indeed, given the choice, why would a researcher intentionally select a time of exposure that presented a *greater* challenge to his or her hypothesis? Again, theory led the way, in this case by determining the time of exposure. What at first glance seemed to be a straightforward method, no more than a matter of counting grains, turned out to be mutable, a matter of choice or inclination.

There was one final manipulation of the data and it was the *coup de grâce*. It ensured the desired outcome. As in the cell fractionation studies, certain locations were excluded as a source of radioactivity *a priori*. Grains found over them were either declared artifacts or were subtracted as background noise. In this way, a natural provenance was attributed to some structures and discounted for others. And this determination was made *in accordance with the dictates of the vesicle model*, not nature. Yet again, assumptions controlled results, and unsurprisingly, the cytoplasm was center stage. Grains found over it were considered background noise to either be ignored or subtracted.

In any event, what at first appeared to be a clear-cut experiment, almost pure observation, turned out to be far from it, and in many cases was little more than choosing one's method to produce a preconceived result. Years later, a high-resolution method of identification was developed in which tiny gold beads were conjugated to individual protein molecules.

This overcame many of the difficulties of resolution with autoradiography. Now all that was needed to identify the location of the protein was to find the minute gold beads in the electron micrograph.

But this changed little. A wonderful example of how little is seen in the first study published using this method of identification on the pancreas. The observations led to head-scratching conclusions. Beads were not found in the cisterns of the ER where they were expected, while at the same time, they were found over the cytoplasm where they were not expected.

What to make of this "incongruity?" Not to worry. Their absence over the ER was an artifact, the result of their loss during sample preparation, while their *presence* in the cytoplasmic space was an artifact of redistribution due to their release from the selfsame ER during sample preparation. In other words, what was observed was an artifact, and what was imagined was true. According to the theory, the beads should have been present where they were not and should have been absent where they were present. And so in an act of tortured reasoning, it was concluded that what was actually observed had it completely backwards. The truth was found not in what was observed, but in what was imagined. The gold beads had greatly improved resolution, but nothing else.

One day, a colleague at The University of California, San Francisco (UCSF) stopped me in the hallway and introduced himself. He said furtively, or so it seemed, that he had something he wanted to show me and asked if I would come to his office. As I sat down in the chair next to his desk in the dimly lit room, he opened the bottom drawer and took out a pile of photographs. They were autoradiographs of pancreatic tissue. The pictures surprised even me. Under the bright light on his desk, I could see the grains, and there were a great many, lying unambiguously over the cytoplasm. Indeed, this was the dominant feature of the photographs.

He went on to tell me about his experiments and under what circumstances he had obtained this result, but when I excitedly asked for the citation to the published work, he said that he had not published it. When I asked why, he replied that he probably could not get it published, but even if he could, he feared retribution. He was afraid that he would be excluded from national meetings or even more worrisome would lose grant support. He was just not willing to take the chance.[10–12]

## chapter eleven

# Location, location, location!
## Where am I?

> What a piece of work is a man! How noble in reason, how infinite in faculty! In form and moving how express and admirable! In action how like an Angel! In apprehension how like a god! The beauty of the world! The paragon of animals! And yet to me, what is this quintessence of dust?
>
> **William Shakespeare**
> *Hamlet, Act II, Scene 2*

You may have noticed that most of the assumptions that we have been talking about are assumptions of place, of location. It goes without saying that if you know nothing else, to know that something is moved from one place to another, you must know that it is *in* the place from which it is thought to come and that it ends up *in* the place where it is thought to go. Conversely, we can know that movement does *not* occur from or to a particular place if whatever is being moved is *not* in it. This is awfully obvious, and it is what the experiments we have been talking about were presumably designed to find out. Was the labeled protein here or was it there, was it here at one time and there at another?

Against this, we have learned that neither the location nor the path traveled was actually determined by measurement, but instead by a series of assumptions. As explained, this led to experimental results that were preordained; that were known in advance. With assumptions posing as facts, the theory validated itself. Whatever one thinks of the vesicle theory, this is what happened.

The problem begins at the beginning, and by beginning, I mean with the initial disposition of newly manufactured protein. Where does it go? The vesicle theory tells us that all proteins manufactured on ribosomes attached to the endoplasmic reticulum (ER) are transferred directly into its interior or cisternal spaces and that having entered them are held there, sequestered or isolated, and unable to leave without the assistance of vesicles.

If this was *not* true, if this was not the fate of some or all of these proteins, then they would end up in the forbidden cytoplasm and, by default,
would be able to move freely about the cell without the help of vesicles.
Similarly, if they had entered the cisterns, but could cross back into the
cytoplasm, they too would be able to move around the cell without the
aid of vesicle mechanisms. In either case, once in the cytoplasm, whether
to be secreted or to enter other cellular compartments, individual protein
molecules would have the opportunity to pass into and through intervening membranes to reach their destination.

Thus, substantiation of the vesicle theory depended critically on proof
that each and every protein molecule made on ribosomes attached to the
ER was directly and immediately transferred into its cisterns by means
of some universal irreversible transport process. That this is, in fact, what
happens was and remains widely believed. Evidence for the belief rests on
the results of two experiments that were carried out many years ago. The
first, performed about a decade before the second, supplied physical and
quantitative proof of entry, while the second was chemical and qualitative
and was intended to expose the mechanism of transfer as well as demonstrate its irreversibility.

But in fact, neither study actually measured the entry of proteins
into the intact ER of living cells. In its place, uptake was measured into
microsomes, or at least that was the claim. I say "the claim" because once
again comforting assumptions took the place of the actual physical determination of location. Remember that microsomes are fragments of the
endoplasmic reticulum produced during tissue homogenization. Let me
explain.

To start, it was presumed without proof that microsomes were sealed,
membrane-enclosed objects that could serve as surrogates for the intact
ER. In the first study, new protein was labeled as it was being manufactured on isolated microsomes. The microsomes were then sedimented by
centrifugation and resuspended in a label-free medium that contained
a detergent to disperse the enclosing membrane and release the microsome's contents.[13,14] The labeled protein found in the medium after this
treatment was assumed to have come from within the demolished microsomes, having of course been transferred into them in the first place.

Where else, it was thought, could it have come from? There was no
other possibility, was there? But in fact there were *three*. Material adsorbed
(weakly attached) to the surface of the microsomal membrane, weakly
attached to the surface of the ribosomes themselves, and most importantly suspended in the extramicrosomal space of the sediment, the large
volume of fluid outside the microsomes. It would also be released under
these circumstances. Whatever was thought of the potential magnitude
of these other sources they were ignored. They were just *not* taken into

account; their potential contribution was never assessed. As a result, the claim that the labeled protein came exclusively or even mainly from inside the microsomes was no more than a reassuring guess.

The second study was even more suspect.[15,16] In it, no pretense was made of physically establishing location. Instead, a chemical difference between two molecules was presumed to specify their location. Of course, this was silly. *By itself,* unadorned by any assumptions, a chemical difference between two molecules can only be proof of that difference. It tells us nothing about their physical location. Perhaps they are in my room or on the moon. How would I know? Indeed, perhaps one is in my room and the other on the moon.

The chemical difference itself was completely unexpected. It turned out that some proteins manufactured on ribosomes attached to microsomes lack a particular peptide segment, or sequence of amino acids, that was present when the same protein was made on free ribosomes, on ribosomes not attached to microsomal membrane. The most straightforward explanation for this peculiarity was that the missing peptide was produced in one case and not the other.

But however straightforward, this possibility was not considered. Instead, it was proposed that the peptide *was produced in both situations* and removed or deleted when it was made on microsomes. More than that, this deletion was said to occur pursuant to the protein passing into the internal spaces of the microsomes (once again assuming the presence of an isolated space). Critically, the absence of the extra peptide was said to be proof that the protein was inside the microsome. That is to say, its location could be inferred from its chemistry.

However, there was the niggling problem of proof. A variety of rather obvious pieces of evidence were missing. First, it had to be shown that the extra peptide segment was indeed present on the protein chain as it was being made on microsomes for at least a brief period of time before it was removed. Second, the segment had to be recovered separate from the parent molecule. Third, a one-to-one correspondence had to be established between the two.

Next, there was the issue of location. It had to be shown that the shortened protein and its deleted segment were actually *in* the microsome (as well as in the intact ER in the living cell), and nowhere else. Finally, all this had to be known quantitatively, it had to be shown for each and every protein molecule made on attached ribosomes. In fact *none* of this was known at the time of the proposal and, to the best of my knowledge, remains unknown all these years later. While all this proposing was going on, a young researcher in my laboratory, Jenny Ho, decided to carry out a simple "control" experiment with a surprising—one might say *scandalous*—result.[17]

Having produced a labeled protein in the pancreas in situ, she removed the gland from the animal, homogenized it, and collected microsomes by centrifugation. She then asked two basic questions. First, was the labeled protein inside the microsomes where it was thought to be, and second, if it was to what could its presence be attributed? To answer these questions, she made two measurements.

The first was ridiculously, even irritatingly, self-evident. She washed the microsomal sediment in a large volume of fresh medium (in the absence of a detergent) to remove the labeled material adsorbed to their surface as well as labeled material suspended in the medium. This minimal procedure, one wash, reduced the labeled protein content of the microsomal sediment by a whopping 75%. That is, at least three-fourths of the labeled protein associated with the sediment was *not* in the microsomes, but on their surface or in the surrounding medium (or if inside, leaked out with the wash). With additional washes, this number rose to over 90%.

In the second measurement, a large molecule was used as a marker. A labeled form of the polysaccharide inulin (not insulin, but inulin, a large carbohydrate molecule that is composed of many sugar molecules joined together) that did not normally enter cells was added to the medium just *before* the tissue was homogenized. This was done to see how much of the marker would be trapped in the microsomes as they formed during homogenization. This would tell us what percentage of the labeled protein found there was an artifactual *inclusion* from other sources in the cell that took place during homogenization. The result was astonishing. The two numbers were essentially identical. That is, the percentage of inulin trapped and the percentage of the labeled protein found were indistinguishable. This meant that *none* of the labeled protein could be definitively attributed to the cisternal spaces of the intact ER *prior to homogenization*. It might all be an artifact.

Why, we asked ourselves, hadn't similar measurements been made in these other studies? Their absence was a serious fault. Science cannot manufacture nature to its liking. It cannot say that the location of something is known without actually knowing it. I believed this was a demonstration of the vesicle theory's infection, mortal infection, with whatever assumptions were needed to verify the model. Perhaps the most remarkable thing about the whole business was that *the absence of something could be seen as proof of its existence*. This was the case for the rare omega figure and the invisible budding microvesicles. Their presence was assumed to be too fleeting to be seen. Here, the absence of something (a peptide) was seen as proof of its existence. How does one argue against such intangible proof?

Along with a variety of subsidiary assumptions (some of which are outlined in chapter twelve), these studies formed the basis for a model known as the "signal hypothesis." With great specificity and a complexity comparable to that of the parent vesicle theory itself, the signal hypothesis laid out the events thought to account for the first step in the vesicle process, the transfer of protein into the ER.

As it was originally imagined, the signal hypothesis proposed that the missing peptide segments were located at the leading edge (the N-terminus) of the elongating chain of amino acids as the protein was being synthesized (remember proteins are chains of amino acids). This forward-most segment was called the signal peptide, and it was thought to serve both informational (signaling) and mechanistic functions. Its presence *signaled* that the particular protein being made was to cross the membrane enclosing the ER to enter its internal space. At the same time, its presence *enabled* passage. During their transfer, enzymes embedded in or associated with the membrane removed or deleted the peptide from the growing chain, thereby sealing the protein's fate. Pursuant to its removal, the new protein could no longer cross membranes. It was now trapped in the ER waiting to be rescued by vesicle mechanisms.

Though called the signal *hypothesis*, it was about as far away from a simple testable hypothesis as one can get. Rather, it was a complex series of propositions and assumptions, along with supporting evidence. It was many hypotheses rolled into one grand concoction. As a result, discordant observations did not lead to its rejection, but to its elaboration. If a new wrinkle was needed, it was added, invariably making things more complex. To accommodate observation, the theory was altered but never challenged. The signal hypothesis, like the parent vesicle theory, was perfectly elastic; its central premises remained intact no matter what.

For example, when it was found that some *secreted* proteins *lacked the signal sequence* and yet passed through and out of the cell, this did not lead to the rejection of the signal hypothesis as one might innocently expect, or even rejection of its claim of universality. Instead, it was said that the signal peptide *was indeed present* in these proteins, but just not at the beginning of the peptide chain. There was an *internal* signal that served the same purpose. Eventually, it became clear that *no* particular sequence of amino acids was the signal sequence (as was originally envisioned). In fact, no common sequence of amino acids or even a clear, shared structural or functional theme was ever established.

Yet, none of this seemed to matter. And it did not take long before the signal hypothesis became the signal *mechanism*. Inconsistent observations seemed to add new detail, new knowledge, and indeed new proof of its workings. It was Ptolemaic science at its best, endlessly modifiable, and never challenged by experimental test. The signal hypothesis was a master example of the new biology of the cell and its vesicle paradigm.

I will have more to say about this in the next chapter, but before we move on, I would like to change the subject slightly and consider the relationship between hypotheses and evidence in general. Towards the end of my career at the University of California, San Francisco (UCSF), I taught an elective course for graduate students on the scientific method. Most of the students who took the course thought that it was going to be about making accurate and reproducible measurements, about making sure to use trustworthy techniques and appropriate statistical analysis when carrying out experiments. They were wrong.

While critical to good science, this was *not* the scientific method I had in mind. My purpose was to expose the students to the logical and philosophical basis of scientific knowledge—to the real scientific method. I was particularly interested in the distinction between evidence and interpretation, and given the discussion thus far I hope you can see why.

This distinction seemed particularly easy to make in experimental biology, because it was the common practice of journals to have separate "methods," "results," and "discussion" sections. The first was to describe the methods used, the second to present the data and only the data, and the third to discuss and analyze it. The editors had smoothed the path for the reader.

Regrettably, however helpful in theory, in practice these neat barriers were breached as often as they were honored. Most important, interpretations of experimental results often appeared unannounced in the results section, intruding on the straightforward presentation of observations. Sometimes this was inadvertent or unavoidable, but it was also a subtle way of investing the interpretation of experiments with the imprimatur of results. In such cases, the interpretation appeared to be what had actually been observed. In some papers, the two sections were merged into a common "Results and Discussion" section, leaving it to the reader to figure out which was which.

When I told the students that their assignment was to distinguish evidence from interpretation in a particular piece of work, they thought that it would be laughably easy. After all, they were smart, and in many cases, the editors with their section headings had paved the way even for the densest among them. Much mortification followed when, in paper after paper, they conflated the two. They had had no trouble *defining* the difference in the abstract, but when it came to discerning it in a particular piece of work, they often found it hard to sort out. The two were often fused in their minds, as they followed various traps laid by the authors.

Each semester we considered different papers, all of them chosen by the students, except for one that I chose. My choice was the same every year. It was the article (actually two contiguous articles) that presented chemical proof for the signal hypothesis. Invariably in examining these papers, students failed to distinguish the evidence for the authors' claim from the claim itself. What had been proposed seemed to prove its occurrence. The hypothesis served as its own evidence. What was interesting was that despite the students' initial confusion, when I pointed out the distinction between what had actually been observed and what was conjecture, they grasped it immediately. Once the blinkers were removed, what had previously been so hard to see now seemed pretty obvious.

## chapter twelve

# Making the impossible real
## Could the impossible be possible?

> Reasoning will never make a man correct an ill
> opinion, which by reasoning he never acquired.
>
> **Jonathan Swift**
> *Letter to a young clergyman, 1719*

In spite of the many weaknesses in the evidence for the vesicle theory, indeed absent any experimental evidence for it whatsoever, there was an enormously powerful reason for its acceptance. This was based on two commanding understandings about the biological cell at the time of the theory's development. The first was that the biological membrane was a cellophane-like envelope of lipid molecules occasionally pierced by tiny pores through which water and other small molecules could pass. The second was that protein molecules were far too large to fit through the tiny pores and that, as water-soluble substances, they were unable to dissolve in or otherwise pass across the rigid lipid membrane.

These beliefs, essentially universal among biologists at the time and foundational to their understanding of the character of the biological cell, were based on contemporary physical and chemical knowledge of protein and membrane structure. Together, they led to the seemingly irrefutable conclusion that biological membranes were *impermeable* to protein molecules as a class. These large water-soluble substances simply could not pass through them.

Since many proteins moved from place to place within and out of the cell, and in the process crossed both intracellular and cellular membranes, some other mechanism had to account for their passage, and as explained, the only alternative was a vesicle mechanism. Yet there was something strange about this conclusion. Even though it seemed undeniable at the time, this was not advertised in talks or in print as a compelling argument for the vesicle theory. Why the aversion?

There was a reason and it was also compelling, but in a very different way. The conclusion about the membrane and its impermeability to protein molecules shed unwanted light on a fundamental contradiction in the vesicle theory. It had been determined that proteins were manufactured on ribosomes, and ribosomes were either free in the cytoplasm or attached to the external surface of the membranes of the endoplasmic reticulum (ER), and critically, only their external surface. There were no ribosomes *in* the ER, attached to the internal face of its membrane or simply floating in its cisterns.

This made no sense. If new proteins were in the cisterns as the vesicle theory told us and, at the same time biological membranes were impermeable to them, then they had to be manufactured *inside* the ER. How else could they get there? Ribosomes *had* to be inside the ER and yet they were not. This either meant that the new proteins were *not* in the cisterns, and that would herald the end of the vesicle theory, or that the membranes of the ER were permeable to them, and that would make a central argument for the theory moot. Either way, there was a serious problem.

There was a solution and it was not merely distasteful, it was unpalatable. The sole way to keep the proteins out of the cytoplasm and in the cisterns was *if each and every protein molecule manufactured on attached ribosomes crossed the membrane of the ER.* In other words, the vesicle theory seemed to require what it claimed was impossible—a membrane permeability to protein molecules. What an extraordinary contradiction!

We have just talked about the scheme invented to resolve the contradiction—the signal hypothesis. It was a masterful feat of jujitsu, imagining a permeability of membranes to protein molecules that was so restrictive that it could be fairly said that it both existed and did not exist. It envisioned membranes being permeable to proteins while simultaneously claiming that they were impermeable to them. Remarkably, permeability was proof of impermeability. But however clever, the signal hypothesis did not really resolve the contradiction. It merely changed its terms and locution. It restated it.

The unique restrictions that the signal hypothesis placed on permeability for proteins and the intricate mechanisms it required to achieve the obverse objectives of permeability and impermeability are fascinating and can be seen both in what was proposed and what was prohibited:

1. Protein permeability is limited to the membranes of the ER. It is found nowhere else. (At one point, there was an attempt to extend the concept to other membranes, such as bacterial membranes, but this was quickly discarded).
2. Protein molecules can only pass through this membrane *once* in their life history—during their synthesis.
3. Synthesis begins on ribosomes free in the cytoplasm for all proteins, but for those destined for the cisterns of the ER, the ribosome attaches to the external surface of the ER membrane, synthesis

pausing *en route* to prevent the emerging protein chain from entering the cytoplasm.

4. This "docking" activates mechanisms that make the previously impervious membrane, permeable to proteins.
5. This is accomplished by the formation of a small pore or pore-like structure in the membrane.
6. As the chain of amino acids elongates, it passes through this newly formed pore in snake-like or linear fashion.
7. The presence of the signal sequence prevents the chain from curling up into a ball or some other similar three-dimensional structure that would be water-soluble and hence prohibit passage across the membrane.
8. The nascent protein chain can only cross this membrane in one direction, from ribosomes to cisterns.
9. The possibility of return is prevented when the signal sequence is cleaved during passage.
10. As a result of its cleavage, the chain curls up, curves, bends, twists, crimps, and twines as it enters the cisterns, eventually nestling into a unique and thermodynamically comfortable three-dimensional, usually more or less spherically shape with a polar surface and a nonpolar core *that cannot cross membranes*.
11. The protein is thus trapped in the cistern's watery environment.
12. When the completed molecule is released into the cisterns, the ribosome is released from the membrane.
13. With its detachment, the pore disappears.
14. The membrane once again becomes impenetrable to protein molecules.
15. The new protein, now sequestered, is unable leave the cisterns without the aid of vesicles.

What an extraordinary series of ideas. The inherent contradiction of permeability *qua* impermeability was resolved by proposing that only certain proteins can cross membranes, only at one location (the ER), only as they are being made, only in an elongated (linear) form, only in one direction, and only once in the protein's life history. Yes, the membrane of the ER is permeable to proteins, but only in the most restricted fashion imaginable. It is the remarkable exception that proves the rule.

And so, quite amazingly, what is not possible takes place and yet remains impossible. The whole business, with its restrictions and intricacies, was based on the phantom signal sequence, whose absence signified its presence. If you think about all this for a minute, you will realize that nature demanded none of it. The signal hypothesis fulfilled a great need all right, but that need was born of the vesicle theory and then current views about membrane permeability to protein molecules, not nature. It satisfied man's ideas, not nature's needs. But of course, it is not nature's task to be of service to theories of man's design. *The theory must fit nature, not the other way around.*

This raised an interesting question: Of what benefit were the mechanisms outlined in the signal hypothesis to nature? Why would such a complex mechanism that limited the passage of protein molecules to one membrane, in one direction, on only one occasion in its life history have evolved, other than to meet the needs of the vesicle theory and its preconceptions? An affirmative answer does not jump to mind.

Still, and even though the signal hypothesis was only able to resolve one contradiction by introducing another, the concept was dazzling. The membrane was permeable, *and* the vesicle theory was saved at the same time. The cloud of assumptions and complexities on which the vesicle theory had been floating were simultaneously abrogated and confirmed. And so it came as no surprise that despite its tortured nature, the signal hypothesis received a warm welcome. Whatever the actual facts, it had saved the day. It saved the vesicle theory from its fundamental contradiction. How absolutely stunning!

Still, there were two huge difficulties. First, it had become clear that neither the character of the membrane barrier nor the properties of proteins were as I have described them. Indeed, what was believed to be true was mostly false. The membrane was neither a cellophane-like wrapping, nor were the proteins simply large inert polar spheres. There will be more about this later.

The second difficulty is the core subject of this book. While the *belief* that membranes were impermeable to proteins was strong and entrenched, *it had never been experimentally verified.* There was no *evidence* that membranes were actually impermeable to protein molecules. The conviction was so strong that measurements seemed unnecessary. Indeed, attempting to measure what did not exist would not only be an exercise in futility, but would also be unscientific.[18]

# *Nature's way*

## chapter thirteen

# Testing the theory
## Testing the already known

> Falsehood flies, and truth comes limping after it, so
> that when men come to be undeceived, it is too late;
> the jest is over, and the tale hath had its effect.

**Jonathan Swift**
*The Art of Political Lying, The Examiner, #14, 1710*

As a young scientist trying to make sense of the vesicle theory, of what
was believed and why, I found that the flimsiness of the evidence supporting the theory was not accompanied by uncertainty about its authority,
but to the contrary by enormous confidence in it. And this was not merely
the attitude of those who developed the theory, but the whole field of cell
biology.

As best I could figure out, there were three reasons for this. First and
foremost, there was faith in the appropriateness of the assumptions we
have been discussing. Second, there was trust in the say-so and expertise
of the scientists who carried out the research. And finally, there was a
kind of breathlessness at the sophistication of their methods. I found each
of these troubling. I was also disturbed by the unpleasant suspicion that
some of the evidence was specious, that it had been selected, manipulated,
or misrepresented.

Surely, I thought, science was *not* about some sort of categorical acceptance of a particular set of beliefs, akin to the faith of a religion. It was
about questioning belief, not submission to it. Indeed, it seemed to me
that this was what it meant to be a scientist. As a matter of intellectual and
professional pride, scientists would bend their beliefs to accommodate
nature's facts. After all, they would say, theories are just theories.

As I began to question the vesicle theory in public and in print, I
learned just how mistaken I was. My view of science and its purveyors
was an idealization, a fairytale. I was utterly unaware of how passionate
and fierce belief in scientific theories can be, how unyielding and uncom-
promising. I soon found out that one could not question the vesicle theory
without the questioner, not just the question, being attacked. Both to my
face and behind my back, with overt, shouted hostility, or soft, cunning

snideness reeking of intellectual sophistication, I was variously called disagreeable, offensive, uninformed, ignorant, noncomprehending, obsessed, and finally unhinged.

I was completely unprepared for the barrage. Naïve and oblivious, I was in a haze of idealistic self-delusion. As the theory failed various tests, I credulously expected others to join me in questioning its notions. Not only didn't this happen, there was no accommodation whatsoever. What I thought was convincing evidence against the theory was not met with excitement or at least interest, but with anger, rejection, and derision. I found myself under attack for expressing doubt. What gall; what cheek; what chutzpah!

Even those who seemed to agree with me, who saw the weaknesses and understood that our observations presented a real challenge to the theory, usually expressed their opinions under their breath, sometimes literally whispering in my ear. In public they were mute, cowed, or cowardly. From time to time, I was offered friendly advice. Why, they would ask, was I causing such a ruckus? What was the point? Attitudes were not going to change. I was not going to convince anyone. Do something else with your time!

Over the years, the vesicle theory failed dozens of tests in our hands. Sadly, however undeniable the results, and whether they came from one test or many, they failed abysmally to pierce the armor of a world that seemed to be populated by closed or fearful minds. At the time, I wondered whether this was my fault. After all, I was young and inexperienced. Maybe there was a proper way to question a theory, with a certain delicacy and sensitivity, whose refinements I had not yet learned. Certainly, the attacks on me were not sugarcoated.

At any rate, whatever the cogency of our experiments, however gripping the results and whatever the combined impact, they were greeted with anger, rejection, and derision, by fear and cowardice and by committed attempts explain them away. It was said that our technique was faulty, that our results were artifacts of one sort or another, and if neither criticism sufficed, then our observations were most certainly properly explained (more like explained away) in a fashion that was agreeable to the vesicle theory. *It* accounted for what we saw, our claims to the contrary notwithstanding.

Even if all this failed, the unnerving results could always be ignored, put out of one's mind, out of the collective mind of the field. Just written off. They were simply not worthy of note. In any event, our whole body of work was either spurned or disregarded. It had no serious impact on the beliefs and attitudes of cell biologists. Still, these wonderful experiments shaped my scientific odyssey. Benighted or not, they are at the heart of this book. We will get to them in a minute, but before we do, we need to make one final preparative diversion.

To understand why things happened the way they did, to understand why our studies were treated so dismissively, we have to talk a bit more about the scientific method. I raised its great shadow in the *Praeludium* as well as from place to place in the discussion to this point. This may elicit yawns of boredom from some of you. Get on with it! But this is no mere scholastic exercise. Having a sense of science's method is critical to any consideration of our experiments and the subject more generally, not to mention scientific knowledge more broadly. What was our purpose, our scientific purpose, in performing the experiments? What were we after? And what were the motives of those who greeted them so contemptuously?

To answer these questions, we have to look beyond the attitudes of the individuals involved in the dispute to the views of the 20th century's two most influential thinkers about the nature of science, the philosopher Karl Popper and the social commentator, historian, and physicist Thomas Kuhn. Two schools of thought emerged from their ideas, not surprisingly called "Popperian" and "Kuhnian." Their apostles have in one way or another been at each other's throats for decades because their perspectives have been thought to provide opposing views of science.

I did not think so. In my mind, the two schools of thought were perfectly compatible. They just focused on different aspects of the same activity. They seemed to be arguing past each other. Popper was concerned with the theory of science and Kuhn with its practice. Of course, both must be taken into account in any attempt to explain the enterprise.

My intentions in carrying out the experiments (I shall discuss some of them in the following chapters) are found in Popper's point of view and the motivations of my adversaries' in Kuhn's. Befitting a philosopher, Popper understood science in formal terms. He said that true science had two characteristic features: deductive tests of theories and the rule of falsification. The proper scientist, as opposed to the mere researcher, formulated hypotheses deduced from some general theory as a means of testing that theory's soundness. Does nature behave as the hypothesis, and hence the overarching theory predict? If it does, and if such and such should happen and it does, we say that the theory is affirmed, shown to be correct. If does not, then the hypothesis is false and by extension so is the larger theory from which it is derived.

Popper tells us that if we cannot devise deductive tests of a theory, then it is not a scientific theory, however sophisticated it may seem otherwise. The reason is simple. If we are unable to test our theories, then we have no way of knowing whether or not they are true or false, correct or incorrect. All we have, all we can have are prejudices, our own biased counsel to guide us in the search for the truth.

Popper's second principle, the principle of falsification, conveys the depressing news that all science can truly, that is, unambiguously know about nature is what is not true, what is false. Tests that affirm our theories, that suggest that they are correct, and that thereby give them credence, necessarily leave their actual truth or falsity eternally contingent. However seemingly incontrovertible and however numerous the affirmations, scientific theories are by their very nature always vulnerable to being proven false by the next test.

Hard-won scientific understanding does not become self-evident at some point, beyond question, as a result of experimental or theoretical success. Doubt is always and forever legitimate. Not only that, but to be genuinely dedicated to science, to truly be a scientist, the investigator must try to show that scientific ideas, theories and models that are thought to be true, including his or her own, are *false*. Only ideas that are exposed time and again to the fire of doubt, to tests that may show that they are false, but have survived the conflagration, have a legitimate claim to be descriptive of nature's properties. In this process, skepticism is an act of the highest sort—the truest, most sincere, most faithful, and ultimately, the most insightful kind of scientific thinking.

Kuhn's view was very different. Grounded in history, it was less a matter of theory than practice. It also had two elements. The first concerns what constitutes scientific belief and the second how that belief changes. According to Kuhn, at any given time, scientists in a particular field of study hold a fixed set of beliefs in common that save for relatively unimportant details are inclusive of all the properties, mechanisms, and processes known in that field of study. Kuhn called such belief systems "paradigms," a term he borrowed from the social sciences and that has long since entered our everyday vocabulary. A paradigm is an exemplar or archetype. It is the way something is thought about, a set of beliefs.

Scientific paradigms embrace a multiplicity of things. Naturally they include theories, as well as the evidence and methods used to support and develop them. But in addition they include all sorts of suppositions, presumptions, and assumptions, supported and unsupported, substantiated and unsubstantiated, as well as prejudices and biases of all kinds, including lies. To this ungodly mess we must add the political and social context in which the beliefs are acquired and held. This includes the education of students, claims of authority, expressions of power and position, access to funds, as well as scientific misdeeds, such as the distortion and selection of data, even outright fabrication. All of it, from theory, to education, to evidence, to assumptions, to data selection and fabrication, to the power of authority, are devoted to one goal and one goal alone—*promoting* the perspective of the paradigm.

This takes us about as far away from Popper's simple testable theory as possible. In the world of the paradigm, the mandates are wholly different. Scientists who labor within its borders, in what Kuhn calls "normal science," seek evidence to strengthen it, to elucidate and clarify its claims, but, and this is critical, not to test or question them. Its constituent models are understood to be true in general, to provide a proper description of nature's properties. At any given time, the paradigm is thought to provide their more or less complete characterization in the particular field of study. Whatever doubts and uncertainties remain are either inconsequential or solvable within the borders of the paradigm.

Kuhn believed that the great majority of scientists work in the world of normal science. Their task is to fill in gaps in knowledge, to determine what remains unknown or uncertain *in a system that is thought to be true as a general matter.* Their research is like striving to complete a jigsaw puzzle whose solution is known from the picture on the box top, except that in this case, the task is not to realize a picture, but to perfect the viewpoint of the paradigm.

Despite the alignment of powerful forces devoted to strengthen the paradigm, to make it permanent, Kuhn knew as a matter of history that there are no fixed beliefs in science. From time to time, indeed time and again paradigms are upended. And yet, he wondered, how could this happen? How could such a change take place if all the experts in an area held the same beliefs and were working assiduously toward their perfection?

Who would rebel, who would break the bond of allegiance? In other words, who would become a traitor to the common cause? Kuhn answered this question by introducing a kind of *deus ex machina*, someone from a different discipline, who, ignorant of the mandates and constraints of the belief system, "upsets the apple cart" and puts the paradigm in jeopardy with a new piece of evidence or a new theoretical insight. Whether intentional or not, the outsider introduces conflict into a pacific world of comfortable and shared understanding.

As Kuhn described it, when this happens, followers of the paradigm do not, as one might innocently expect (as I did), respond by accommodating their views to the new information. Instead, they rush headlong to the barricades to defend the *status quo ante* and do so with great ferocity. The challenge or challenger is deemed unworthy. The demurring scientist is dishonest, incompetent, and in any event, the paradigm can explain it all, or at least everything worth explaining, at least well enough. In response, the upstart may gather further evidence, posit further arguments, highlight previously obscure flaws in the paradigm, thereby instigating new attacks on the bastion. A battle for primacy ensues that Kuhn called a "scientific revolution."

In this view, changes in scientific understanding do *not* occur gradually, a little bit at a time, a little here and a little there, as scientists slowly hone in on the truth of the matter as is often thought. This gradualist view of scientific progress dominated thinking for centuries and is credited to Francis Bacon in 16–17th century England. Kuhn disagreed wholeheartedly. Instead, he saw a cataclysm in which the old paradigm is toppled, and a new order, a new understanding, a new paradigm is installed. Popper's agreeable experimental tests and deductive logic are *not* the change agents in the real world of science, in Kuhn's world. Things there are not so pure or so rational. Belief in the old paradigm, in the old view, was not disinterested, but committed, colored by strong biases, and often by harsh, even rotten social and political forces both within and outside the academy.

As a young scientist, I was a committed Popperian, naive about the world of Kuhnian debauchery I was about to enter. In varying measure, humans share two opposite traits of character—skepticism and naivety. I was at one and the same time a skeptic about scientific knowledge *and* a naive Popperian. My skepticism was expressed in my belief in experimental test to assess the validity of theories, and my naivety in the belief that the results of such tests were the standard bearers of scientific progress. I was in an amiable Popperian reverie from which I was about to experience an abrupt awakening to the ugly Kuhnian world.

With these thoughts in mind, we are now ready to tell the story of *Proteins Crossing Membranes*. As I have explained, only with a veritable phalanx of preconceptions, did the experiments we have discussed provide validation of the vesicle theory. Most important among them was the understanding that all proteins made on ribosomes attached to the endoplasmic reticulum were *excluded* from the cytoplasm. Yet, and this was my driving motivation, against this deeply rooted mindset, with its many assumptions, presumptions, and suppositions, stood one simple fact—*Despite assertions to the contrary, the actual properties of protein movement in the living cell were unknown!*

What was thought was in great part ersatz. It took place in a simulacrum, in a model of human invention, not the natural world. It was the properties of the artifice, of the model, not nature, that had been studied. In the mixed up world of this new cell biology, nature and models of nature had become fused. They had become one and the same. When, as a young scientist all those years ago, I first read the experiments said to provide the evidence for the vesicle theory, I was flabbergasted. What others saw as illumination, I saw as divination.

The evidence, what there was of it, was to my untutored (read: unindoctrinated) mind, very weak. As said, most importantly, there was little to no observational basis in the living cell to support the complex model, with its many parts and elements. If this was not a sufficient cause for skepticism, the alternative to a vesicle system—protein movement via the cytoplasm with proteins crossing membranes on their own—had been excluded not by experimental observation, but by *a priori* assumptions. This left the vesicle theory without competition, the guaranteed victor in a one-horse race.

Still, at first I was perfectly comfortable with the vesicle model. I believed that it was probably true despite the deficiencies. Indeed, I thought that if fortune shined on me, I would be the person who provided the missing proof, who would overcome the deficiencies, who would show that things happened just as the vesicle theory proposed. Thus, I began as an acolyte. However, it did not take long before it became clear that I was not suited for the role. I found it impossible to merely accept what others had claimed. I was a skeptic by nature, and as it turned out by profession as well.

Crucially, however widely heralded and redolent the data and ideas of the vesicle theory, I found it impossible to impute physiological meaning to them. To me, the theory was not much more than an elaborate guess. Assumptions masking as fact had taken the place of deductive logic as the tool of choice. My desire in the face of this situation was to carry out experiments that sought to reintroduce deductive logic and its tests to the study of the process. By testing the vesicle theory, we would establish the actual properties of protein movement, particularly those involved in secretion, as they occurred in the living cell. What did the cells of the pancreas actually do as they went about their business?

Our first observations did *not* affirm the vesicle theory. Though my initial inclination was to question our observations, not the model, before long, and despite great discomfort, I found that I just could not accept the vesicle theory and decided to carry out additional tests, hoping that the results would reassure me about its validity. But they did not. Quite the opposite, they reinforced my suspicions. As the list of failed tests grew, I became increasingly disillusioned with high-flown pronouncements about the vesicle theory, and became a full-fledged skeptic.

The controversy that resulted, some 50 years in the making now, was extremely one-sided. I had no real allies in the fight, and to many, I was an iniquitous character. Though isolated and alone, without friends and allies, without a community of support, I pressed on in this lonely effort. I was sustained and comforted by those who shared the struggle with me—the students, fellows, and technicians who worked with me, and most importantly, my wife and family. And then, there was the comfort of the amazing observations themselves, beautiful evocations of nature. They were uplifting, a source of inspiration, and provided the impetus to press on despite the obstacles.

Out of naivety and idealism, or perhaps, less kindly, stubbornness and idiocy, I had the temerity to question what had become a time-honored scientific belief. A belief that today is so firmly ensconced that it is taught as truth to most middle school students. Add the indelible fact that the Nobel Prize was awarded for the vesicle theory having been proven true in 1974, and on two occasions since, 25 years later in 1999, and most recently in 2013, and you can see that being a skeptic was fraught.

Whether as my younger self or in my present incarnation as an ancient curmudgeon, those who know of the controversy often see me as a devil or a saint, infamous or courageous. I am neither. In telling my story and that of my students and colleagues, I am not interested in either defending my reputation or in changing fixed attitudes in cell biology. I am power-less to do either. I am interested in something altogether different.

As already discussed, as a college student I was drawn to science by what seemed to be a miracle. Humans had this wondrous capacity. We were able to ask questions of nature and obtain meaningful answers. I learned that even I, as an ignorant undergraduate, possessed this gift. Above all else, this book is about this miracle, about the wonder and excitement of scientific discovery.

If I could magically redo my young self with the wisdom and judg-ment of age, perhaps I could have navigated the turbulent waters more skillfully. As a young scientist, I did not understand that belief trumps all, in science as in life, even tests of scientific theories. But of course, the past is the past and the young must be allowed their ignorance, because from it comes the wonders of unexpected discovery.

As Kuhn explained, naivety often enters the world of settled belief, the world of the paradigm, in the guise of the outsider. The naïf is either unaware or insufficiently informed about the rules of the game and goes where those who know better dare not go. What took place in my case was a combination of the skepticism of Popper's world of tests and the naivety of the outsider ignorant of the demands of the paradigm, of enshrined belief in the Kuhnian realm.[19,20]

As said, our observations were either dismissed or ignored. And when it was not possible to do either, our interpretation was nonetheless still seen as wrongheaded often without clear reason and always without serious evidence. It was stated, with all attendant authority, that when properly explained, our results fit within the borders of the paradigm. There was no contradiction. We will see how all this took place in the chapters that follow.

In them, I describe particular observations or series of connected observations chosen from a larger set that were tests of the vesicle theory. I have two reasons for my choices. First, they are illustrative and second, (I hope) not too difficult to understand. In my descriptions, I have tried to stick to concepts and avoid unnecessary technical detail. Unfortunately, if you have read this far, you may appreciate that it is not possible to do this completely and yet present a more or less accurate picture of things. At the end of the book, I give references to source material, including some of the primary research publications in which the experiments discussed were first presented. The data and issues have to one degree or another been considered in two previous books. The first, published in 1985, was technical. The other, published almost 20 years later, was part of a discussion of broader issues about reductionism and the scientific method. Citations to these books and to various review articles on the subject are listed at the end of the book.

## chapter fourteen

# Ring the tocsin
## Into the fray

Over the past half-century research on protein secretion in eukaryotes has in great part been devoted to accumulating evidence for the vesicle theory and its various subsumed mechanistic claims. This research though voluminous has almost without exception failed to test the theory and these claims in their own right, against the properties of the natural processes they profess to explain, or against mechanistic alternatives, believing such tests and comparisons to be unnecessary. And though there has been some modest acquiescence to the presence of a non-vesicular route of secretion..., widespread opinion today is that the vesicle model has been experimentally validated even if all of the details have yet to be worked out. That is, we can have confidence that the mechanisms it envisions provide a realistic description of the chief agency, and for many the only agency, of protein secretion in eukaryotic cells.

Yet, in spite of this broad consensus, there is a range of evidence at odds with the vesicle theory, some of it unmistakably falsifying.... The first explicitly and openly contradictory evidence was published more than 30 years ago ... and in the years since scores of observations inconsistent with the theory have been reported .... Naturally, in raising questions about the vesicle theory, a non-vesicular alternative was implied. But many of these observations did more than imply the existence of such an alternative; they revealed its nature and explanatory power, and in so doing, brought the vesicle theory's generating understanding and raison d'être that membranes are impermeable to protein molecules,

into question. It was discovered that just like small
molecules, proteins are able to pass through intact
membranes by means of various membrane embedded transport mechanisms.... Furthermore, and
remarkably, non-vesicular processes appeared to
account for protein secretion to the exclusion of vesicle mechanisms. And yet, in great part because the
vesicle mechanism of secretion was thought to have
been established, this evidence in all its variety, however contradictory, compelling, or falsifying, and
though provoking questions about the vesicle theory that called for convincing counter-evidence, not
just rationalizations, has by and large been ignored
or dismissed. It and its associated ideas have been
regarded as controversial, rather than illuminating,
and however wrongly, adjudged false, artifactual,
misinterpretations, oddities, or minor and irrelevant discrepancies to an established understanding.

**From Stephen Rothman**
*The Incoherence of the Vesicle Theory of*
*Protein Secretion, Journal of Theoretical*
*Biology 245: 150–160, 2007*

It was a beautiful autumn day, cool and sunny, and I had just come down
from Boston to present my recent experiments to his group at Rockefeller.
*He* was the father of the vesicle theory. His Nobel Prize was still about
5 years off, but the stars were aligned, and he was being feted as a leading
light of the new cell biology. He entered the small seminar room just as
I was about to begin my talk, tall and swarthy in a dark suit with a dark
tie, and a full head of straight, slicked back black hair. He was an imposing figure, but also an ominous one—alluring, yet alarming—positively
Transylvanian. And then there was the rumor that he had been a physician in the infamous Romanian medical corps during World War II.

But my apprehension as a young scientist did not arise either from
admiration or his potential approbation or disapproval. What worried me
was that once I described my experiments, gave away its secrets, his large,
well-oiled research group would steal my thunder. They would coopt
what I had discovered and it would become theirs. My small laboratory,
with one technician and a couple of students, just could not compete. How
badly mistaken I was would become apparent before the seminar was
over.

As discussed, the high-resolution beam of the electron microscope had
exposed a whole new world of anatomy within cells. Biological cells were

jam-packed with all sorts of tiny structures, many seen for the first time, and others only guessed at in the fog of the traditional light microscope. Given the importance of these discoveries and the new biology they promised, it seemed inevitable, perhaps long overdue by the late 1960s, for the Nobel committee to award its prize to a scientist or group of scientists who had played a formative role in this unearthing, and he was a prime candidate.

His prominence however was not the result of his contributions to the development of the complex and demanding methods required to examine cells in the vacuum of the powerful electron beam, but to his vesicle theory that assigned a central role to some of the newly discovered structures in the critical activity of moving proteins from place to place within and out of the cell. It was one thing to develop a method, even to see new things, but quite another to grasp the significance of what you saw, and this was what he and his colleagues had done.

My seminar was about experiments that I had just published in *Nature* magazine, which I hoped would resolve the dispute discussed earlier between Pavlov and his student Babkin.[21] As it happened, these experiments also provided the first test of the vesicle theory against the properties of intact secretion cells. This had prompted the invitation.

My observations were straightforward. The relative rates of secretion of two digestive enzymes were measured before and after the administration of a hormone that increased the rate of enzyme secretion by the pancreas. I had discovered that the hormone changed the proportions of the two enzymes in the secreted fluid. Thus, secretion was *nonparallel*, confirming Pavlov's point of view and undermining Babkin's. As for the vesicle theory, it came down on Babkin's side of the controversy.

Quite simply, the vesicle theory and its exocytosis mechanism proposed that the various proteins were released from the cell *en masse*, without favor or preference. As a result, their proportions in secretion would match those in the cell and its secretion granules at the time. In other words, they were released at commensurate rates or as Babkin put it "in parallel." And so the simple observation that on occasion this was not the case, that secretion was nonparallel or that it differentiated among the proteins that were secreted raised worrisome questions about all the wonderful imaginings of vesicles moving to and fro, fusing with and separating from membranes. As such, they had landed with a thud at Rockefeller.

In my talk, I described the observations, the methods I used to make them and the conclusions I had drawn from the data to the small group of people who filled the cramped room sitting around a table in the center and standing against its walls. As I was going through the facts and figures of the experiments, projecting slides onto a small screen in the corner of the darkened room, there was an audible murmur, a kind of groaning from the audience. At one point, the fellow sitting next to the head of the laboratory bent over and whispered something in his ear. He nodded approvingly. When the talk ended and the floor was opened for questions, I learned what the muttering was about, and why my fears were misplaced.

Much to my surprise, *none* of the questions concerned the challenge that the results presented to the vesicle theory, or how the theory had to be modified in light of them. They were about one thing and one thing only—the methods I had used to measure the two enzymes. I had described them in some detail, but my explanation had not satisfied them. They wanted to know more. With each question I went over things again, expanding upon and clarifying one fine point or another.

But nothing I said seemed to mollify them. They wanted to know how I could be sure that my measurements were correct? Though they did not point to any specific problem or missing controls, my explanations, though detailed, had left them unconvinced. They wondered whether the change in proportions I reported might just be a measurement error, an artifact? If they had been properly made, would the proportions have remained unchanged in line with Babkin's point of view? I explained why this was not possible. But I was not able to assuage their suspicions.

They were not questioning my conclusions *if* my measurements were correct; they were questioning the measurements themselves. Could I be *absolutely* sure that there had not been some sort of measurement error? Under the pressure of questioning, I conceded that though I had no idea of what that error could be, I could not eliminate *all* possible doubt. How could I, how could anyone? With this admission, there was a collective sigh of relief.

This was what they wanted—an acknowledgment of the *possibility* of error! It gave them permission to say to each other and subsequently to others with knowing nods and raised eyebrows that my results were *in all likelihood* the result of a measurement error, and should be viewed skeptically. "In all likelihood" meant that the null hypothesis (that there was no change) applied, that Babkin was correct, and so was the vesicle theory. By the time the seminar ended, and without so much as a fare-thee-well, "could" had morphed into "were." My measurements *were* in error, and so, thankfully, the vesicle theory—not to mention the hoped-for prize—was not at risk.

On the flight back to Boston, I thought about what had happened. My first reaction was to go back to the lab and redo the measurements. But I knew that the numbers were not going to change. It did not matter how many times I repeated them or in what way I altered the protocol, the result would be the same. The values were true and accurate. To search for phantom errors in need of correction would be a fool's errand. Where would I look, anyway? I realized that even if I could magically achieve absolute certainty, or find the elusive error and correct it, it would not per-suade them unless they got the result they wanted. They would remain unpersuaded, disinclined to accept what I reported, *unless* it supported their theory.

Rather than waste my time trying to convince people who could not be convinced, rather than going for a ride on a merry-go-round to nowhere, I decided to expand the circle of inquiry. I would examine the phenomenon of nonparallel secretion in situations beyond those of the initial report. I was especially interested in Pavlov's ideas about the regulation of diges-tion. In addition, we would perform tests of the vesicle theory that did *not* depend on the relative rates of secretion of different enzymes and that thus would not be open to the charge of measurement artifact.

Whatever I decided to do, others could of course attempt to corrobo-rate my results. Corroboration, after all, is the widely regarded and reas-suring security blanket of science. If my observations were false, then others would be unable to reproduce them. If they were correct, they would confirm them. As it turned out, though straightforward in prin-ciple, reproducing my observation turned out to be far easier said than done.

There were two major barriers. One was methodological; the other, psychological. The methodological obstacle was general and applied to many biological studies. Their complexity and the large number of vari-ables involved, often made faithful reproduction very challenging, and my experiments were hardly an exception. To be successful, one had to pay very close attention to each and every detail of the experiment you were attempting to repeat. Nothing could be assumed to be irrelevant.

As for the psychological issues, those who sought to reproduce my observations were, in the main, not disinterested observers, but devotees of the vesicle theory. They believed that my observations were wrong and *expected* to be unable to corroborate them. They had a vested interest in a negative result. To expect such individuals to scrupulously reproduce the original observations would be delusional.

Their inability might not be due to outright dishonesty, but to more subtle forms of bias. For instance, if they were skeptical of the methods of measurement that produced the original observations, they would think that rigorous fidelity to them would be foolish. They would want to apply more satisfactory methods that would, in their view, make the effect clearer if it were real.

Anyway, the thinking went, if it were real, the phenomenon should be robust enough to be observed in the face of small differences in approach and method. Petty variations should be immaterial. As things trans-pired, these supposed trivialities included a long list: the species studied, making measurements in situ, rather than in vitro in the organ culture system in which the original observations were made, measuring different enzymes or, as said, measuring the same enzymes in different ways, applying different stimuli or different doses of the same stimulus, studying things over different periods of time, at different times of day, in fed not fasted animals, and on and on. As said, the thinking was that if the phenomenon was real, such relatively unimportant differences should not matter.

However, if corroboration is the goal, even seemingly inconsequential differences can be of great, if unanticipated significance. From time to time, Ph.D. students complained to me that all their experiments produced negative results. Regardless of what they did, they saw no difference between treatments. While no effect is a frequent outcome of scientific studies, their inability to see differences was often of their own making. I would find that they had not paid sufficient attention to the intimate details of their experiments. They would do it one way on one day, and another on the next. I would try to impress upon them (indeed I tried to enforce the principle) that, having settled on an experimental protocol, they had to stick with it or start all over again. This did not mean they had to rigidly apply it and never change anything, come hell or high water, but they had to understand that changing procedures in midstream, however seemingly immaterial, without clear and imposing reasons for doing so would in all likelihood increase the variability of what they observed and might, as a result, leave them with, as they complained, everything looking much the same.

The greater the variability between the results of individual measurements and individual experiments, the *less likely* a statistically meaningful difference will be found even if it exists, indeed even if it is substantial. In other words, the greater the variability, the less likely that it will be possible to reject the null hypothesis that there is no effect.

In regard to the observation of nonparallel secretion, observers with the preconceived notion that no change in proportions should be seen had no incentive to keep the variability low. In fact, it was their friend and ally—the more, the merrier. The expectation of a negative result encouraged a kind of insouciance and sloppiness in trying to repeat the experiment. Given this mindset, finding nothing was not only easy; it was satisfying, even gratifying. And, as explained, it is far easier to see nothing, than something. It was fine not to see an effect, because this was just what they anticipated.

Several studies published in the years following the first report of nonparallel secretion claimed to be failed attempts at corroboration, as indeed they were. But they not only failed to reproduce my result, they failed to reproduce my experiment. More about this later, but first let me tell you a sad tale about the seductive nature of collaborations designed to resolve scientific disputes. The collaboration in question was about nonparallel secretion and was intended to either corroborate or refute its existence.

Government granting agencies and the like are often enamored by such collaborations. They think that goodwill, if not among men, then at least among scientists with different opinions, will lead to the resolution of disputes based on the observed fact pattern. In the case of nonparallel secretion, it would either be corroborated or not, and whatever their inclinations, the parties would be forced to agree. What better use of money for scientific research than to resolve a significant scientific dispute?

Such collaborations can work when the dispute is minor and there is a general agreement among the parties about the broader perspective. When, and this was the crux of the matter here, broader questions are at issue they tend to be recipes for disaster. This was a most personal disaster for one of my students. His thesis work had been path-breaking. We will talk about his experiments in the next chapter. He had recently won first prize for the best graduate student research in the nation at the annual meeting of the Federation of American Societies of Experimental Biology (FASEB), at the time the flagship organization for experimental biology worldwide.

As a result of this spectacular start, he was offered a tenure-track Assistant Professorship at Cornell University Medical School, and along with it, a laboratory in which to begin an independent research career even before he received his Ph.D. degree. He was not going to have to suffer the postdoc miseries and then struggle to find a position. Here it was on a silver platter. Not only that, Cornell's medical school was not in Ithaca, by Cayuga's waters, but in Manhattan. This meant that he could move his family back to their native and beloved New York City. It was a wonderful opportunity, or so it seemed.

But there was an ominous sign. Cornell's medical school was located across the street from the Rockefeller Institute (now the Rockefeller University)! This worried me, but how could I tell him not to accept such a good offer? My concern turned out to be warranted. The appointment came with a catch. Rather than continuing the line of research he had pioneered, which had been so successful and had so much promise, he was pressured into collaborating with a member of the vesicle group at Rockefeller to determine whether my observation of nonparallel secretion was correct.

Coming from opposite sides of the dispute, the two investigators would once and for all resolve the controversy. I warned against the project. He should pursue his own research and not get inveigled into a collaborative project with members of the vesicle group. They had an agenda and would use him. But what would his bosses think if he were unwilling to engage in this collaboration? He would appear defensive and paranoid. After all, the potential collaborator was an honest man and they were both seeking the same thing, the truth, weren't they? The power of their results would force agreement and the controversy would be resolved.

What followed was predictable. The collaboration was a debacle. But far more important, it was a tragedy for this exceptional young scientist and his family. The two men agreed about nothing. They were hardly able to design experiments, much less carry them out. The effort went on for a few years and failed to produce any meaningful results. If you believed that the original observations were fallacious, this was not a bad outcome. You could say that despite all your efforts, it had been impossible to reproduce the phenomenon, and that would certainly be true however disingenuous.

Having been distracted by this wild goose chase, my ex-student soon found himself facing the tenure clock with little to show for his effort. Before long, he lost his position. Only after many years of struggle was he able to find another university position with a laboratory in which to do research. Even then, he was never able to continue his promising research on the mechanisms of secretion.

Indeed, obtaining funding for anything that questioned the vesicle theory had become virtually impossible. Not a single graduate student or postdoctoral fellow of mine was ever able to continue or, better yet, develop the research they started in my laboratory however successful it was and however much they wanted to. It was hard enough to get an academic position; no less obtain the required government research grant, but to get a position and obtain a grant to carry out studies that raised questions about the vesicle theory was a certain recipe for academic and scientific failure.

Still, despite all this, the original observations of nonparallel secretion were a worry for vesicle theory enthusiasts. They were like a low-grade fever that just would not go away. In this regard, something unexpected took place almost 10 years after the first description of phenomenon. It revealed that despite the passage of years, nonparallel secretion remained an irritant. I was to give a talk about the results of a new test of the vesicle theory as part of a symposium at the annual FASEB meeting.

It was the largest gathering of biologists in the world at the time, and attracted 25,000–30,000 scientists to the dilapidated, yet-to-be-rebuilt Atlantic City and its boardwalk. Scientists could be seen wandering to and fro on the boardwalk, talking to each other, mumbling to themselves, and rushing between the large run-down "grand" hotels and the convention center to catch one talk or another.

My talk took place in a ballroom at one of the larger hotels. Hundreds of seats had been set up in orderly rows. Upfront was a stage with the largest screen that I had ever seen. Standing next to it, almost swallowed up in its immenseness, was the podium. The talk was well attended, with the overflow leaning against the walls.

I thought that the observations I was going to discuss were powerful and would convince anyone with a mind to be convinced that the vesicle theory was misconceived. I started with a quote from Karl Popper about testing theories to set the stage for what was to come. I laid out the experiments, explained how they were done, and what I believed they demonstrated. When I finished, the floor was opened for questions and I was about to experience a bad case of déjà vu. It was as if I was back at Rockefeller all those years ago. The first question had nothing to do with what I had just spent almost an hour talking about. Instead, it was about the validity of the original report of nonparallel secretion.

By this time, I thought that the issue had been resolved and resolved affirmatively. Whatever the lot of my original report there were some 50 published examples of the phenomenon, for widely different situations, from different laboratories, in different species, for different enzymes, in response to different stimuli. It stretched credulity to think that they were all artifacts of measurement. Not only that, but in the intervening years, we had shown that the phenomenon was grounded in Pavlov's ideas about the regulation of digestion.

It should have been unmistakable to anyone other than a committed vesicle theory devotee that nonparallel secretion existed. But the questioner was such a devotee. As far as he was concerned, the critique of the vesicle theory stood or fell on the original observation of nonparallel secretion. Nothing else mattered, and he was going to show those assembled that it was in error.

And so, he began not with a question, but a denunciation. "Your observations of nonparallel secretion were mistaken." He then asked the projectionist to put a slide up on the screen. "This attempt (unpublished results not by the questioner, but by someone else who was not specified) to reproduce your results, failed. Look, the ratio of the two enzymes did not change. There is no such thing as nonparallel secretion."

Had a legitimate attempt at corroboration really failed? Though many features of the experiment differed, for instance, they were carried out in situ, in anesthetized animals, not in the in vitro system I had used, it employed the same stimulant, at the same dose, and they measured the secretion of the same two enzymes over roughly the same period of time. I was having a hard time seeing the data on the giant screen, so, I came down off the stage and walked down the center aisle towards the back of the auditorium to see things more clearly.

What the questioner said was true: the ratio of the two enzymes had indeed remained unchanged despite the administration of the hormone. Nonparallel secretion was not seen. But as I looked at the data, I noticed something peculiar. The response to the stimulant was surprisingly small. It had only increased the rate of secretion by about 50%, while I had reported increases of an order of magnitude or more for the same dose of hormone under what were presumably less favorable *in vitro* circumstances. For whatever reason, their experimental system had responded poorly to the stimulant. I pointed this out and said that the effect was so small that the experiment in essence compared controls to controls. They saw no change, because they had produced a negligible effect.

The data had been presented in the common form as averages and deviations from the mean (the variability) for each enzyme at each time point. The greatest source of variation in any experiment that compares the rates of secretion of different substances are differences in the *level* of secretion from trial to trial, from animal to animal. Though this variation can be very great, it can also be easily factored out. This can be done by plotting the individual data points (for the secretion of one molecule versus that of another) rather than compiling them into averages. Displayed in this way, the data form a line, with its slope being the ratio of the two enzymes over the range of levels of secretion.

Changes in its slope would indicate changes in the proportions of the molecules being secreted. There was another related potential source of variation. The effect of the stimulant could vary with the *magnitude of the response* as well as the dose. In this case, the data would form a curve, not a line, with the distribution of data points reflecting variations in the ratio with variations in the magnitude of response.

This turned out to be the case. In response to the administration of this particular hormone, the rates of secretion of the two proteins formed a sharply pitched curve that depended on the magnitude of the response. There was little change in the proportions for data close to the origin, for low responses, as in the data he presented, but when the response was robust, as in our study, a substantial change in proportions was observed.

In any event, for the questioner, the original observation of nonparallel secretion was the lodestone. If it was in error, if it was wrong, then there was no such thing as nonparallel secretion however many other examples of the phenomenon there were. And if nonparallel secretion did not exist, then the whole critique of the vesicle theory was mistaken, whatever other evidence might be marshaled in its behalf. In this light, it was no surprise that in his comments that day, he did not even note the subject of my talk.

Despite what by that time was my longstanding refusal to attempt to reproduce my original observation of nonparallel secretion, seeing it as a pointless waste of time and effort, his question prompted me to do just that more than 10 years after the original publication. If others had failed, would I also fail? Could I corroborate my own results? Using the same system we used originally, and with appropriate care and attention to detail, we not only succeeded in reproducing it, the effect was even clearer, even more pronounced. There it was, for all to see, one more time.[22]

Yet, as the questioner at the FASEB meeting made clear, there were those that still doubted the existence of nonparallel secretion. I would bet dollars to donuts that if you asked vesicle theory adherents today, many if not most would either be unaware of the phenomenon or would believe that it had been debunked long ago. If they were more knowledgeable, they might say that, yes, nonparallel secretion occurs, but this has little to no bearing on the validity of the vesicle theory. It is an irrelevant asterisk that can be explained in terms that are compatible with it.

Given its existence, there were two possible explanations for the phenomenon consistent with the vesicle theory. First, nonparallel secretion might be the result of differences in the rates of *synthesis*, not the transport or movement of the enzymes. Synthesis, not transport, was nonparallel. The second was that the enzymes were segregated in different granules that were subsequently released from the cell by their selective exocytosis. Though the common opinion at the time was that this was not the case and that exocytosis was random, if it was true, it could be said that the results showed that Babkin was wrong, that selective secretion occurred, but that at the same time, the vesicle theory was right. Nonparallel secretion would be the result of selective *vesicle* mechanisms.

Of course it is one thing to make claims, but facts are something else. In light of the facts, neither explanation made sense. For the circumstances of the study, it took just a few minutes to see nonparallel secretion, while it took a few hours to see more than trace amounts of newly manufactured protein in secretion. As for the segregation of different enzymes in different granules and their selective exocytosis, there was no evidence that such a complex system existed. Years later, it was shown that various digestive enzymes (conjugated to gold-labeled antibodies) were found cozily housed *together* in the same granule.[23]

At any rate, when all was said and done, like it or not, the conclusion just could not be avoided. Nonparallel secretion occurred and could *not* be accounted for by vesicle transport mechanisms. The original critique not only remained valid, it had become far more prepossessing, more compelling, supported by more evidence however one might want to wish it

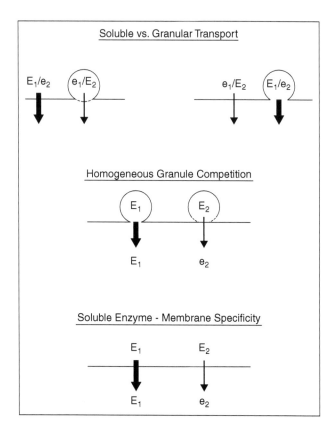

*Figure 14.1* The fateful inclusion. In the original description of nonparallel secretion, I included the possibility that it was the result of the movement of the digestive enzymes across the cell membrane from the cytoplasm ("Soluble vs. Granular Transport" or "Soluble enzyme—membrane specificity"). This was controversial because this was thought to be impossible. (From Rothman, S.S. "Non–parallel transport" of enzyme protein by the pancreas. *Nature*, 213:460–462, 1967.)

away. Nonparallel secretion was neither an artifact nor a curiosity, and, however distasteful, presented a serious challenge to the vesicle theory (Figure 14.1).[24,25]

## chapter fifteen

# The clarion call

## The impossible takes place

> The history of intellectual growth and discovery
> clearly demonstrates the need for unfettered free-
> dom, the right to think the unthinkable, discuss the
> unmentionable, and challenge the unchallengeable.
>
> **C. Vann Woodward**
> *Report of the Committee on Freedom*
> *of Expression at Yale, 1974*

The phenomenon of nonparallel secretion had brought two foundational beliefs of the new biology of the cell into question. The first was that a mechanochemical system of vesicle transport accounted for protein movement between compartments within the cell and to the outside. The second was that protein molecules were unable to pass through biological membranes.

The latter understanding had been unequivocal. It seemed absolutely certain that proteins could not penetrate the membrane barrier, and the digestive enzymes made by the pancreas were no exception. As already noted, based on then current knowledge of the structure of biological membranes and proteins, there was great confidence that this was true. In fact, the living cell seemed to depend on this restriction. The loss or leakage of the cell's life-giving proteins was a sign of pathology that if left unchecked would lead inevitably and inexorably to cell death. The two, life and protein permeability, appeared mutually exclusive.

Given the firmness of this belief, it was no surprise that the inability of proteins to cross membranes was understood to be true *a priori*, known in advance without the bother or benefit of measurement. Though making such measurements was straightforward enough, it would, as said, be a foolish waste of time, a quest to establish the existence of the nonexistent. Not only that, but it would disown a central belief of biology and, as such, would be an apostasy, a renunciation. Consequently, even the attempt bordered on the impermissible. With such a severe attitude, it was not surprising that it had *never* actually been demonstrated that membranes were impermeable to any protein molecule, no less to protein molecules as a class. Belief had trumped knowledge and in no uncertain terms.

The seemingly unassuming phenomenon of nonparallel secretion had changed all that. It said that the impossible might occur, and that some biological membranes might be permeable to some proteins. This put me in a very awkward position. Not to measure the permeability of pancreatic membranes to digestive enzymes would not only be to accept the assumption of impenetrability on its face, but would also be to acquiesce to the judgment that nonparallel secretion was either an artifact or in some unidentified and unproven way the result of vesicle mechanisms. More generally, and more importantly, it would say that being *forbidden* was a proper *scientific* reason for demurral. It would say that a theoretical prohibition was reason enough to not determine facts, even though experimental science forbids nothing that can be measured.

On the other hand, if we had the temerity to make the measurements and by chance found that membranes were permeable to proteins, we would have opened a Pandora's box of speculations, of uncomfortable questions about membranes, proteins, and cells, that not only went beyond the pancreas and its digestive enzymes, but beyond the acceptable. It would be to reject what had presumably been established. If the impossible occurred, then the ideas of biologists about the properties of membranes, proteins, and cells were mistaken in fundamental ways (see chapter 16).

Yet despite this awkward situation, nonparallel secretion was a clarion call. Were the membranes of the acinar cell permeable to digestive enzymes? The question demanded an answer. And so with much trepidation and despite an enormous psychological barrier, we decided to make the taboo measurements. Did these proteins cross the relevant membranes or was their passage in fact prohibited?

The graduate student whose distressing story I told in the last chapter, Charles "Chuck" Liebow, took on the unnerving challenge. If the membranes were impermeable, it would set us free. We would be absolved. The questioning and the controversy would end. There would be proof positive that the vesicle theory was correct. How else could movement occur? On the other hand, if the membranes were permeable, the vesicle theory as a general and universal mechanism to account for the passage of protein molecules across membranes was false.

With a mixture of excitement and apprehension, Chuck began making the forbidden measurements starting with the simplest case, the zymogen granule. Was its membrane permeable to the proteins it stored? The experiment was shamefully easy. One could collect large numbers of these objects almost pure as white sediments from homogenates of pancreatic tissue. All one had to do then was suspend the sediment in fresh medium and look for the appearance of the protein contents of the granules in it.

Did the granules steadfastly hold on to the digestive enzymes or did they release them? Not only were they released, but the extent of release varied in direct proportion to the volume of medium in which the granules were suspended. If the volume was small, most of the contents remained in the granules. However, as the volume was increased, more and more granule content was found in the medium. In fact, one could vary the contents of the granules from full to empty in a predictable fashion simply by adjusting the volume of medium in which they were floating.

Beyond that, for any given volume, the concentration of protein in the medium achieved a stable or stationary value. This was exactly what would be expected if protein movement across granule membrane occurred in *both* directions. With time movement in opposite directions would come to equal each other, that is to say, the system would equilibrate. To show that this was indeed what happened, Chuck performed experiments in which the volume of the suspending medium was kept constant while its protein contents were removed by filtration to continuously renew the gradient from granule to medium. If a concentration gradient of protein was indeed responsible for release, discharge from the granules would continue unabated until they were emptied despite the constant volume and this is just what happened.

All this was just as the law of diffusion predicted.[26,27] Based on these and other observations, we concluded that the granule membrane was permeable to the proteins it contained. Movement occurred in and out across its enclosing membrane in response to concentration gradients. Well then, it seemed that at least some biological membranes were permeable to at least some proteins, and the presumption that membranes posed a universally impenetrable barrier to this class of molecules was dead wrong! In regard to the vesicle theory, it meant that the *cytoplasm* was both a source and destination for proteins stored in the granules. The permeability of the granule membrane to the proteins it stored had completely upended the vesicle theory's *raison d'être*.

As exciting as these observations were, it came as no surprise that vesicle theory adherents unhesitatingly discounted them. They did not dispute that release occurred. Indeed, they saw it. But (surprise, surprise), they said that it was an artifact. A permeable membrane did not exist. Release was the consequence of something called "lysis."

Remember that in the cell fractionation studies discussed earlier, a significant portion of the labeled protein was found in the supernatant or nonsedimentable fraction of the homogenate. This was explained as an artifact of inclusion. That is, during homogenization various natural, membrane-bound cellular compartments broke apart and their contents were released into the suspending fluid. Among these compartments were the zymogen granules. As their membranes split apart, their contents were set free. This was "lysis." Lysis is what takes place when the granule membrane is torn.

It is something like the *hemolysis* of red blood cells. When red blood cells are placed in water, they swell until the cell is so engorged with fluid that its membrane ruptures and its contents, notably the red protein hemoglobin, are released into the medium, leaving behind empty "ghosts" of their former selves. In the case of zymogen granules, the force of disruption was not the entry of water, but friction. As granules rubbed against each other, against other cellular structures and against the glass wall of the containing vessel during the chaos of homogenization, its membrane was torn apart. It was as if a cellophane bag had been ripped open or a glass bottle were fractured by a hammer's blow.

But there was a problem with this explanation, and it was not a minor one. The granules Liebow studied had *survived* homogenization whole and intact. It was *their* membranes that were permeable to proteins. Unsurprisingly, this did not satisfy the critics or diminish the cry of artifact. They insisted that, whatever the circumstances, and for that matter for all imaginable circumstances, whenever protein was released from zymogen granules, it was the result of lysis. This was the only possible explanation because its membranes were impermeable to the proteins they stored, full stop.

Whatever was found, lysis explained it. It was not merely a potential explanation; it was *the* explanation, the only possible explanation. As with the claim of a measurement artifact for nonparallel secretion, *proof* that lysis actually accounted for release was not necessary. The claim was sufficient.

How, we wondered, could we defend our observations against this proposition? How could we ever prove that lysis was *not* responsible for what we found (but see chapter 24)? Lysis seemed to be an all-purpose artifact, always available to explain away distasteful or discordant observations. There seemed to be only one defense against the charge. We had to show that the granule membrane was permeable to digestive enzymes in *intact cells*, where there were no isolated granules to break apart.

And this is what Liebow did. He carried out experiments on intact cells, and the results were absolutely breathtaking.[28] A radioactive digestive enzyme was added to the medium bathing pieces of pancreatic tissue, and its uptake into the tissue and into the relevant subcellular structures was then assessed. Did the cells take up the labeled protein and, if so, did it enter the zymogen granules?

The answer was a resounding "yes" on both accounts! Not only was the labeled material taken up by the cell, it entered the zymogen granules.

However, we found far more than this. The labeled protein had passed through the cytoplasm *en route*. In fact, exchange occurred freely between medium, cytoplasm, and granule—but importantly not the endoplasmic reticulum. Along with the studies on isolated granules, these experiments performed in the early 1970s demonstrated that processes of diffusion, not exocytosis, were responsible for the secretion of digestive enzymes. Towards accomplishing that end, granule and cell membranes were permeable to these molecules. For obvious reasons, we called this the equilibrium model or hypothesis (Figure 15.1).

How would devotees of the vesicle theory react to *these* powerful, hard to refute observations? I was certain that they would discount them, but how? They couldn't point to lysis. There did not appear to be a scientific basis for rejecting them, but nonetheless, there was a way to avoid their acceptance; it was just not scientific.

Politicians use this strategy all the time. If you hold an advantage in an electoral contest, ignore the opposition. Though it was difficult to explain these results in terms that were compatible with the vesicle theory, it was easy enough to ignore them. To the best of my knowledge, the experiments showing permeability in intact cells were never discussed at meetings or cited in print, and they vanished under the heavy, invisible weight of being disregarded.

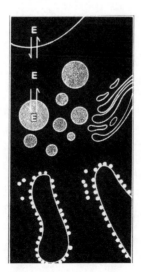

*Figure 15.1* A diagram showing the reversible release of the protein contents of zymogen granules (E) into the cytoplasm and across the cell membrane into the duct (upper left-hand corner).

Orwell would call them "unexperiments:" experiments ignored out of existence. When politicians use this strategy, the idealists among us protest that this is not good for our republic. For democracy to flourish, differences of opinion must be aired, confronted, and not ignored. Otherwise, we run the risk of a society run by self-satisfied oligarchs, oozing noblesse oblige and spouting shibboleths and pieties, absolutely certain of the suitability and rectitude of their beliefs and actions.

If this is bad for politics, it is no less so for science. What kind of science would we have if those with popular views and political power could simply ignore disagreeable facts? Would self-satisfied oligarchs, oozing noblesse oblige and spouting shibboleths and pieties rule science? Would that even be science? In any event, despite being ignored, these experiments provided powerful evidence *against* the vesicle theory. But this is only the beginning of the story.[29,30]

*chapter sixteen*

---

# The membrane revolution
## When is a revolution not a revolution?

> I will look at any additional evidence to confirm the
> opinion to which I have already come.

**Lord Hugh Molson**
*(1903–1991), British Politician*

In the late 1960s and early 1970s, at the same time as the work on zymogen granule permeability was taking place, a revolution in our understanding of the structure of biological membranes was underway. After decades of dominance, the model of a rigid, protein excluding lipid bilayer (a layer of lipid two molecules thick) was losing favor to a new model.

The membrane was still understood to be a lipid bilayer, except that the lipids were not fixed in place as in a cellophane-like envelope. They were in a semiliquid or liquid crystal state and moved rapidly to and fro across and along the plane of the thin membrane they formed. And proteins were no longer excluded. Now they dipped into the lipid layer here and there, and in some cases, transected it entirely. This new model was called the *Fluid Mosaic Model* (Figures 16.1 and 16.2).[31,32]

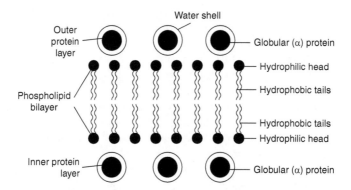

*Figure 16.1* Older (Davson-Danielli) model of the biological membrane. Sometimes called the "Sandwich Model." An external protein layer is shown adsorbed or otherwise bound to the water-seeking (hydrophilic) heads of the lipid bilayer. Proteins are depicted as spheres covered by water.

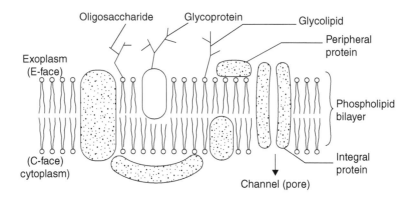

*Figure 16.2* The fluid mosaic model of the biological membrane. Proteins penetrate and cross the lipid bilayer.

The membrane was fluid because of the fluid nature of lipid layer and mosaic due to the assortment of protein molecules dispersed within it. Evidence for the old model that once seemed compelling, now seemed not only weak and unpersuasive, but almost vaporous. Even the idea that the membrane of red blood cells fractured during hemolysis seemed wrong. It had simply been sufficiently stretched by the entering water so that the cell's contents, including hemoglobin, could easily pass through what was now a very porous membrane mesh.

These new ideas excited me, not only in their own right, but also because they gave succor to the idea that biological membranes were permeable to proteins. If proteins were found within the membrane, then it seemed to follow that they could cross it. If they had entered from one side, why not exit from the other? Not only that, but the new model did not require that proteins be lipid soluble to cross the lipid barrier. All they had to do was disrupt this flimsy, mobile lipid layer. The membrane was more like a surface phenomenon, an easily disturbed phase transition (absent an underlying bulk phase), than a cellophane covering.

But I was about to be disabused of such sweet thoughts. Among the inventors of the new membrane concept were two cell biologists at the University of California, San Diego in La Jolla. I invited them to talk about their new ideas at my campus, at the University of California, San Francisco (UCSF), and they graciously reciprocated by inviting me to La Jolla to talk about my research. My experiences with them, one after his seminar at UCSF and the other during mine at La Jolla were enlightening, though not in an encouraging way.

After the seminar at UCSF by the senior member of the pair, we had a little time before I had to take him to the airport for his flight back to

San Diego. We spent it in my office talking about membranes and protein secretion. I was interested in how he saw the connection between the two. How did the new membrane model inform his ideas about the permeability of biological membranes to proteins? With a kind of stridency and defensive certainty that was becoming an old, if unwelcome, friend, he made it clear that the new membrane model did *not* allow for the passage of protein molecules across it any more than the old model. The presence of proteins in membranes, in which he believed, and their passage across them were very different matters. It was one thing for a protein to enter a membrane, but totally different to cross it. Quite simply, the first occurred, and the second was impossible.

I thought that there were two reasons for his insistence, and neither was related to scientific evidence. The first was an attempt to protect his ideas about membranes, and the other was simply a strong belief. As for protection, it seemed that he wanted to shield his new model from anything that might jeopardize its acceptance. It was not yet firmly established, and allowing for the possibility of the prohibited protein permeability could sink the ship. As for the strong belief, as a cell biologist, he was an avid defender of the vesicle theory. He harbored no doubts about it. It was indisputably true.

When I expressed skepticism, particularly noting questions about the presence of digestive enzymes in the cisterns of the endoplasmic reticulum (ER), he became simultaneously agitated and adamant. "What was I saying?" There was, he assured me, compelling, indeed irrefutable evidence that all proteins made on the ribosomes attached to its surface were sequestered in its cisternal spaces. If we had the time (we did), he said, we could go to the library, and he would show me the proof.

When we got there, he began frantically searching the stacks for a particular paper. He was not sure of the journal or year, but eventually, he found what he was looking for. Pleased, he opened the volume to the correct page. Here it was. Here was the compelling evidence, the irrefutable proof! It was a single autoradiograph of the ER with an opaque mass in a cistern with radioactive grains over it. Whatever the opaque material was, wherever it came from, and whatever the source of the grains in the underlying tissue section, such dense masses were *not* common features of the ER. It occurred to me that *this* might be an artifact. In particular, it looked like a piece of a zymogen granule that had been displaced during the preparation of the sample for viewing. But whatever it was, whatever its provenance, this one uncommon image convinced me of nothing. It seemed like weak tea, hardly the irrefutable evidence I had been promised.

Anyway, this father of the fluid mosaic model was no kindred spirit. Despite the radical change he had brought about, his ideas about the physiological character and functional properties of biological membranes had remained unchanged. They were traditional and unaffected by what he had wrought. Most importantly, he continued to believe that membranes were only permeable to small water-soluble molecules. And he believed this with the kind of certainty that by this time was, as said, familiar, but disturbing nonetheless. I could not fathom his closed-mindedness, particularly about membrane permeability to proteins, given the revolutionary character of the model he was so vigorously championing. Even so, he clearly did not like the idea that his model had opened the door to this possibility.

When I asked how he envisioned proteins penetrating the bilayer, he shrugged and said that this was not yet known, but that however it occurred, it did not allow for their passage across it. Some years later, the signal hypothesis provided the explanation he was seeking, proposing that proteins destined for the membrane insinuated themselves into its substance *during* their manufacture. As they elongated on the attached ribosomes, they slithered into the membrane, and once into place, once in position, were stuck there, unable to leave.

At any rate, this father of the membrane revolution was a dyed in the wool believer in the vesicle theory with all its accoutrements. On the way to the airport, trying to be polite and praising my guest, I asked how it felt to be the standard-bearer of a scientific revolution. As with much that I had to say by this point, he took offense. There was no revolution, and he was not its standard-bearer! His discoveries were part of the normal progress of science. His evidence was clear, and others realized it and agreed. As for the vesicle theory, new membrane model or not, it stood strong and unperturbed.

My experience with his younger partner took place a few months later during my talk at La Jolla. Though what happened was different, the message was the same. He had not been at my talk, but as the seminar was about to end, he burst into the large, almost-filled amphitheater through a rear door and sat down in the last row not quietly, but with a ruffle and thump that turned heads. As I was taking questions from the audience, a professor, apparently a microbiologist, asked if I was familiar with the recent research that demonstrated the transport of a protein across a particular bacterial cell membrane *after* its manufacture.

I said that I was and mentioned several other recent examples in bacteria. I then noted that bacteria secrete all sorts of proteins, various toxins, enzymes, and the like, and yet have no vesicles, no exocytosis. With this, the late arriver had had enough. He stood up to issue a pronouncement. He said that while the report in bacteria might be true, it was a peculiarity limited to certain bacteria. But other than such curiosities, there was no doubt that proteins were not able to cross membranes. This was incontrovertible, and he pointed to the vesicle theory for evidence.

I asked, if the proteins were able to enter the lipid layer as he thought, then why couldn't they exit across its opposite face? He said emphatically that they could enter, but with no proof or argument, said just as strongly that they could not cross. This was just not possible. He continued that if this was true anywhere, it was in cells such as those of the acinar cell of the pancreas that I had been studying. Though he had not heard my talk, he knew with a conviction that seemed to border on rage that whatever I had said was wrong.

In the years to come, and not withstanding such attitudes, changes in our understanding of biological membranes led to a new appreciation of their permeability to protein molecules. Most significantly, there was evidence of large membrane pores. Dating back to the early years of the 20th century, it was thought that biological membranes were dotted with small water-filled pores or channels with diameters of a few Angstroms or a few tenths of a nanometer, through which small polar molecules of lesser diameter, such as water, small ions, and small organic substances, could pass. Large molecules like proteins were excluded.

Though widely believed, this conclusion about pore size was not the result of direct measurement. The values were extrapolations. As the size of the molecule being examined increased, permeability decreased *towards*, but not to zero. Pore size estimates were linear extrapolations from the measured permeabilities of molecules of different sizes to the pore size projected for zero permeability.

There were two problems with the extrapolation. First, it provided the average size of the pore. It told us nothing about the distribution of sizes around the mean value. It could not be excluded that there were a small number of pores large enough to accommodate proteins. The second problem was the extrapolation itself. Perhaps, decreases in permeability followed a curve, not a line, that though continuously falling did not actually reach zero. Who could say?

Either way, it could not be excluded that there were a few pores large enough to accommodate protein molecules. If they existed, even if they were few in number, they could account for the transport of a mass of these large molecules that was comparable to the movement of small molecules through a great many small pores. Though as said, the actual permeability of these membranes to large molecules was not measured. And thus, whether they could or could not cross was unknown, and the conclusion that they could not was just a guess, an expectation.

This conclusion of impermeability was reinforced by an unfortunate historical fact. Most early studies on cell permeability were performed on the easy-to-obtain red blood cell. It is hard to imagine a more inauspicious cell to examine if one is interested in the permeability of membranes to proteins. The mature red blood cell is unique in that it does not manufacture proteins and hence cannot afford to lose them, especially its precious cargo hemoglobin. If any cell membrane was likely to be ill-disposed to leak proteins, it was that of the red blood cell.[33]

In any event, not long after the advent of the fluid mosaic model reports appeared of channels in membranes large enough to accommodate protein molecules. Indeed, one could see them. Yes, I said see them. A variety of new methods of microscopy had made it possible to scan the surface of membranes at extremely high resolution. With appropriate techniques, one could even see individual protein molecules, no less large pores or channels. Large pores were most prominent in the nuclear membrane, where they apparently accounted for the critical, abundant, and constant traffic of proteins, as well as large RNA molecules between the nucleus and the cytoplasm.

At about the same time that attitudes about the structure of biological membranes were changing, in the early 1970s, I decided to visit Cambridge, England, and its famous Medical Research Council Laboratory for Molecular Biology (MRC laboratory) to discuss membranes and protein permeability with its denizens. The laboratory was a historic place, home to several Nobel laureates. It was here that the methods of crystal diffraction were first used to determine the structure of proteins. The method had previously been used to determine the three-dimensional structure of small molecules in atomic detail, and now X-ray beams were being used to produce diffractive patterns of protein crystals. Their analysis exposed the three-dimensional structure of these large molecules in the same atomic detail that diffractive patterns specified the structure of small molecules. The exact location of the hundreds of amino acids that comprised each protein chain could now be determined with precision.

In any event, the MRC lab was filled with scientists, young and old, who knew all that was known at the time about the molecular structure of proteins. What did *they* think about protein molecules crossing the lipid bilayer? Large pores aside, could proteins somehow on their own recognizance cross it? Was this impossible as so many thought? I expected these experts on protein structure, the crystallographers, to add their voice to the chorus of naysayers.

To the best of my knowledge, nothing they had found changed the traditional understanding of the physical properties of proteins. If anything, they seemed to confirm it. Proteins were by in large rigid, roughly spherical polar or water-soluble molecules, while biological membranes were mainly comprised of water-avoiding or nonpolar lipids. I assumed that they would tell me, like the cell biologists, that proteins could not cross biological membranes (large channels had yet to be discovered).

So, I was pleasantly surprised to find that most of those I spoke with, including one Nobel Laureate, had no trouble with the idea of the lipid barrier being breached by proteins. From their point of view, who could say what these inordinately complex molecules could or could not do without testing it? It seemed arrogant and presumptuous to conclude *a priori* that the bilayer structure prohibited their passage without checking it out. This was a joy to hear, particularly since we had "checked it out" and knew that protein transport across membranes occurred. To me, the question was not whether this happened, but how? In addition to large pores, was it possible that some proteins could breach the lipid membrane barrier?

In this regard, I was particularly interested in the structures of the two digestive enzymes, trypsinogen and chymotrypsinogen (or trypsin and chymotrypsin in their active forms) that I had studied in the original nonparallel secretion experiments. My host directed me to a storage room that housed molecular models of these and a variety of other proteins. As he pointed the way to the room, he said with a sly smile, "Stephen, look at their surface."

The models were large, extremely complex edifices, much like giant tinker toys made of atoms and bonds. Paying heed to his advice, I began by looking for clues on the surface of the molecules. Both proteins were quite water-soluble, and I expected that accordingly their surface would be covered with polar, water-seeking amino acids.

What I found was a big surprise. I discovered not a single clue, but many. It turned out that, like stars shining brightly in the night sky, about half of the amino acids on or near the surface of both molecules were either nonpolar or amphiphilic (meaning that they could be either polar or nonpolar depending on their environmental circumstances). The surfaces of these proteins were *not* populated solely, or even mainly, by resolutely polar amino acids, with amino acids expressing nonpolar proclivities being buried deep inside to hide from the incompatible water. Instead, here they were on the surface, ready to interact with membrane lipids as well as with nonpolar surfaces of membrane-embedded proteins. At least some of these nonpolar amino acids had been forced to the surface by their attachment to polar neighbors, but whatever the reason for their presence on the surface, there seemed to be far too many to assume that they would have no effect on the physical and chemical properties of these otherwise water-soluble substances.

And then there was the protein crystal itself. For all its atomic detail, the molecular model I was looking at only told us about the structure of proteins constrained in both shape and position by the cage of the crystal. No wonder they were rigid balls. Perhaps freed from this constraint, their three-dimensional structures would be affected by their environment just like other molecules.[34] A few years later, methods were developed that allowed for the examination of the fine structure of protein molecules in liquid media, and we learned that this was indeed the case. Despite their structure being constrained by internal bonds, it changed with the changing environment. It was not fixed, unitary, and immutable, as had been previously thought.

In any event, whatever my meandering thoughts about all this, it was slowly becoming clear that many proteins crossed membranes after their synthesis, that is, post-translationally ("translation" being a synonym for synthesis). Not only did this occur, it was common. In fact, it was more than common; it was essential. For instance, it was needed in order for bacteria to digest food and protect against attack. In other words, it was necessary for nothing less than their survival. Furthermore, neither the nucleus nor mitochondrion of animal cells or chloroplasts in plants could do their jobs without it. Quite simply, without it neither animal nor plant life would exist.

So what was once thought impossible, not only occurred, but was vital. Whatever the mechanisms,[35,36] and many were imagined, it could now be said with confidence that proteins of all kinds passed through biological membranes after their synthesis. Whether passage was *de minimis* and inconsequential or critical to cellular life, membranes seemed permeable to protein molecules as a class!

Coming to this understanding did not take place overnight. It took a few decades, but eventually, this truth came to be generally appreciated. There was one significant exception. Predictably, the new understanding did not extend to the secretion of proteins by plant and animal cells, or for that matter to any protein manufactured on ribosomes attached to the ER. Whatever the case for other proteins, this large class of protein molecules could *not* cross membranes. That was prohibited. They were only able to move around the cell in microvesicle jitneys.

Not only was this belief staunch, but it was also being asserted with increasing conviction. The movement of protein molecules manufactured on attached ribosomes occurred by means of vesicle mechanisms and vesicle mechanisms alone. However compelling, however conclusive our work on the acinar cell of the pancreas was, it was either disregarded or, with time, all but forgotten.

An odd *mental* compartmentalization had taken place. On the one hand, it came to be accepted that some, even many, proteins crossed membranes after their synthesis. And all sorts of experimental evidence demonstrated this. At the same time, it was understood that proteins made on ribosomes attached to the ER, such as the digestive enzymes, could not cross membranes whatever the evidence. What was evidence of permeability for some proteins was impossible for others, whatever the evidence.

*chapter seventeen*

---

# The case of the disappearing granules

## Extraordinary evidence

Extraordinary claims require extraordinary evidence.

**Carl Sagan**
*Cosmos (TV program), Episode 12, 1990*

When all is said and done, in experimental science, it is what is observed that is supposed to matter, isn't it? And if that does not fit with belief, then however popular the belief, it is wrong and must be rejected. It is in this light that the following chapters tell of five failed tests of the final or ultimate step in the putative vesicle process of secretion, exocytosis. These tests were not about whether there is such a thing as exocytosis, or if there is, its role in other processes. Their sole concern was whether or not it accounts for the secretion of digestive enzymes by the acinar cells of the pancreas.

The experiments were intended to test this widely held belief against the actual properties of the secretion process. They were unusual in that they led not to one, but two conclusions or deductions. As with all assessments of scientific theories, they told us whether a particular proposition about nature was true or false. But it was the second deduction that made these tests unusual and uncommonly powerful. In each of the cases we shall discuss, failed tests of a particular proposition not only told us what does not occur, but they also told us what does occur. In learning what was false, we learned what was true. With the rejection of one idea, came the affirmation of another.

The reason for this unique situation lies in the fact that nature only allows two ways for molecules to cross biological membranes. If one is wrong, then by exclusion, the other is right. In particular, either an external mechanism, such as exocytosis, is responsible for passage, or molecules pass directly and individually through the membrane as a result of properties intrinsic to them. The latter mechanisms encompass various expressions of the laws of diffusion that are referred to collectively as "membrane transport."

And so, molecules are either moved or move on their own. If they don't cross the membrane in one way, they cross in the other.[37] As such, the rejection of exocytosis (of being moved) as the mechanism responsible for the secretion of digestive enzymes simultaneously affirms that their secretion is the result of their passage out of the cell through the cell membrane from the cytoplasm into the duct (moving on their own). It was the ability to discriminate between these two modalities that made these experimental tests especially compelling and informative. More than that, by showing what was false, they not only showed what was true, what does occur, but the properties uncovered were those of the affirmed mechanism.

Thus, in the experiments described in the following chapters, nature took its measure of human ideas, rejecting one, certifying another. I have said that it is in judging human theories against the properties of nature that we find the beauty of scientific discovery, and I count the experiments we are about to discuss among them. But as with other experiments about secretion performed in my laboratory, some of which we have already discussed, neither their power nor their beauty mattered in determining what was believed. In this regard, fealty or true devotion to the vesicle theory was a far more commanding force. However dispositive, our experiments changed nothing. Attitudes remained fixed. And as said, today belief in the vesicle theory is not only pretty much universal, it is *de rigueur*, paradigmatic, part of a larger system of belief that is not subject to rejection by evidence alone, or for that matter subject to serious reflection at all. The dogmatism it embodies reared its ugly head many years ago, when the observations we are about to discuss were made, and their plain meaning rejected or ignored. Let's see what it was all about.

"Look." It was a beautiful photograph of the endoplasmic reticulum. There they were, pairs of elongated membranes piled on top of one another, curving gently across the cell. He said, "Look more closely." "I don't see anything." Finally, there it was: his nickname spelled out in the stacks of membranes, as if engraved by nature herself. He wouldn't tell me if this was indeed nature's doing or his artifice. The ambiguity was meant to be instructive. What was real, what was conjured, and who could tell except the observer or the conjurer? I was in his office to talk about a potential collaboration—an experiment to test the vesicle theory.

We had hatched the idea at an academic gathering a while ago. The goal was to put an end to what was by then a burgeoning controversy. The test was simple, the prediction straightforward, and we were both confident that it would end the argument once and for all. Either the vesicle theory was correct or it was not. Either my skepticism was justified or it was not. Either way, we would stand together and show that different points of view in science could be resolved with reason and evidence, not posturing and rancor.

At the time I was a junior member of the physiology faculty, he was a professor of anatomy at Harvard Medical School and a renowned electron microscopist. He was also a vesicle theory enthusiast, as were most microscopists and cell biologists. He had even published experiments that suggested vesicle involvement in the secretion of the tiny hydrogen atom (acid) by the stomach. Yet, he said that his mind was open, and I had no reason to doubt this sweet and gentle man's sincerity. He said that the vesicle theory was just that, a theory, and of course it might be wrong. He believed that the experiment we had settled on would yield a definitive result, and that having been carried out by a supporter of the model and a skeptic, it would carry great weight. We agreed that however things turned out, we would publish our results and state the conclusions forthrightly.

As for the test, over a century earlier, Rudolf Heidenhain found that when enzyme secretion by the pancreas was stimulated, zymogen (secretion) granules seemed to disappear from the acinar cell (Figure 17.1). He drew three conclusions from this observation. The first was that the substances responsible for digestion were stored in the disappearing granules. The second was that the rate at which these substances were secreted was greater than the rate at which granules were replaced and hence the apparent decrease in their number. Finally, and most important to us here, he thought that this reduction was the result of granules popping out of cells, of their actually leaving the cell to enter the duct system. As explained, the modern version of this leaving is the vesicle theory's exocytosis.

Remember that the exocytosis model proposes that granule contents are moved en masse from cell to duct. In its terms, the disappearance of a granule leads *pari passu*, hand-in-hand to the appearance of its contents in secretion. This means that as the granules disappear, the amount secreted would be proportionately increased. Put the other way round, if the disappearance of all of the granules, 100%, led to the secretion of 10 mg of protein, then the disappearance of 50% would lead to the secretion of

*Figure 17.1* Rudolf Heidenhain, a famous 19th century histologist. He said that with secretion zymogen granules are ejected from the cell into the duct.

half as much, 5 mg, and if 25% vanished, the amount secreted would be 2.5 mg, and so forth. Thus, the relationship between the two variables was proportional and linear, and consequently, their ratio is fixed or constant. The more granules lost, proportionately more secretion would take place. And when they were all gone, there would be no secretion.

We agreed that if this relationship was observed, it would provide strong support for the vesicle theory. If not, then exocytosis was not the mechanism of secretion. To test the hypothesis, we used a stimulant that enhanced secretion greatly, producing the almost complete disappearance of granules from the cells of the gland within a few hours. During the process of depletion, we made two measurements: the amount of digestive enzyme secreted and the area of the cell occupied by granules in cells of the same gland at the same time. Was their relationship fixed as the vesicle theory predicted or did it vary?

Against the vesicle theory, we found that it was not fixed or linear. Granule number and the amount secreted were correlated all right, but their relationship was variable, nonlinear. In response to the administration of the stimulant, their ratio varied markedly over time. In particular, the portion of the cell occupied by granules decreased far more rapidly than their contents were secreted. As granules disappeared, much of what they had contained *remained in the cell*, presumably in the cytoplasm, awaiting subsequent secretion across the cell membrane. Even when most of the granules were gone, a significant fraction of their contents had yet to be secreted. In fact, substantial responses to stimulants could be elicited in their absence.

None of this was as the vesicle theory predicted. However, it fit well with the predictions of Liebow's experiments on granule and cell permeability. The results indicated that molecular transport across the cell membrane from the cytoplasm to duct, not exocytosis, accounted for secretion. I prepared a graph showing what we had observed; a curve indicating the variable ratio between the volume of the cell occupied by granules and the amount secreted. Alongside, I plotted the prediction of the vesicle theory, a straight line reflecting a fixed ratio between the two variables. With chart in hand, I walked over to my collaborator's office in a neighboring building to share the results. As I was walking, I was thinking about the paper we would write, and what we would say. But all my imaginings went out the window when I showed him the graph (Figure 17.2).

He agreed with my conclusion. He said that there was no mistaking that the vesicle theory had failed the test. But when I started talking about where we would publish the results, he became extremely uncomfortable. Apologetically and almost parenthetically, he said that though my interpretation was correct, I should not take it to mean that the vesicle theory or the exocytosis hypothesis was false. However provocative, the results simply did not justify such a claim.

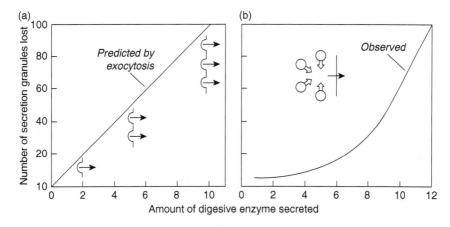

*Figure 17.2* The relationship between the amount secreted and the number of zymogen granules lost as predicted by the vesicle theory (a) and by a membrane transport mechanism (b). (From Rothman, Protein transport by the pancreas. *Science*, 190:747–763, 1975.)

I was speechless. He had agreed that the theory had failed the test, and yet he did not believe that the results justified the claim that it did. What did this mean? I asked if the relationship between the two measures had been that predicted by the vesicle theory, if the ratio was fixed, would that have validated the vesicle theory? Would its claim have been justified? Without a note of irony, he said flatly, "Of course." He continued that the claim, by now "my" claim, that the vesicle theory was false would require far more evidence than this.

I asked if we needed to carry out more experiments? No, there was enough data. The result was clear. Then why, I asked, are you reluctant to draw the conclusion that the data seem to require? He responded sheepishly that to reject the vesicle theory, evidence was needed that was so compelling that criticism of its conclusion was just not possible. In his words what was called for was "extraordinary evidence."

This confused me. I thought that the conclusion was clear and unambiguous, that is, compelling, and he seemed to agree. Wouldn't our data greatly surprise most cell biologists? Weren't they "extraordinary?" Anyway, I continued, when you get down to it, isn't all scientific evidence ordinary? A result may be unexpected, even shocking, but the evidence itself is always ordinary. What else could it be? If we had to wait for some sort of extraordinary evidence before we could reject a theory, we would have to wait a very long time, because there is no such thing. It is the stuff of myths and metaphysics and Public Broadcasting System (PBS) programs, not science. Waiting for it would make the vesicle theory, already widely believed, a fact of nature, independent of evidence and the theory's validity.

He continued that by "extraordinary evidence," he meant facts that would make all knowledgeable cell biologists, or at least those whose opinion mattered, change their minds and reconsider their way of thinking. As we talked, it dawned on me that he was not going to publish these experiments or accede to their conclusion under his name.

Though I was greatly disappointed, when I thought about it, his reaction was not surprising. Though he did not share his reasons with me, despite his well-established status, he might have been fearful of being ostracized. What would others think? How would peer review committees view his future research proposals? His good reputation might suffer if he published the results of our experiments, no less with me. I realized that follow-on research between us was out of the question and decided to publish the results myself.[38]

This experience convinced me of something that we have already talked about. Collaborations between individuals on opposite sides of a weighty controversy, however committed they are to its resolution, however noble their intent, are unrealistic. They dreamily imagine that strong differences of opinion can be resolved rationally just because the issues are scientific. I had slowly come to realize that in the world I inhabited, in the real world of science, such thinking was either a sign of self-deceit or cynicism. Even if both parties were honest and well-intentioned, as I believe my collaborator was, working together towards a common goal when a paradigm or major belief system is being put at risk is almost certain to be unsuccessful. Too much is at stake.

In the years that followed, when I discussed these experiments at seminars, they were invariably met with puzzled silence or querulous responses. Sometimes, an individual would point to other evidence that he or she thought supported the vesicle theory, as if that made our observations and the conclusions drawn from them meaningless. It was as if they lost their significance by means of some sort of special scientific legerdemain. Sometimes, it seemed that all that was needed to successfully disparage the observations was a disapproving glance or a shrug of the shoulders.

Nonetheless, the blunt fact was that no one, at seminars or in print, ever offered a convincing alternative for what we had observed, and I knew of no other explanation, at least no other reasonable explanation. Still and all, content-less denunciation seemed quite enough. Yet regardless of what others thought, the results of these experiments required rejection of the vesicle theory and at the same time affirmed the equilibrium or membrane transport model. As explained, the rejection of one meant the affirmation of the other. There was no way around it.

*chapter eighteen*

---

# What just happened?
## What didn't happen?

> It doesn't matter how beautiful your theory is, it doesn't matter how smart you are. If it doesn't agree with experiment, it's wrong.
>
> **Richard Feynman**

I had recently moved to the San Francisco Bay Area and the University of California, San Francisco (UCSF) from Boston, and research was just beginning in my new laboratory in the modern research tower on Mount Sutro. One morning, Lois Isenman, another recent replant from Boston, whom I had just hired as a lab technician and who would soon become my graduate student, came into my office with obvious trepidation. She was carrying out her first series of experiments and was worried that something was wrong with the data she had collected from the radiation counter.

She had removed the pancreas of a rabbit, placed it in an incubation chamber, and was collecting samples of secretion from its duct. She had added a radioactive amino acid to the medium bathing the gland to make the secreted proteins radioactive. Over time, the new radioactive protein would replace the older unlabeled material in secretion, until every secreted protein was radioactive. This transition would take place regardless of the mechanism of secretion. However, there was an important distinction between a membrane transport and a vesicle or exocytosis mechanism of secretion. According to the vesicle theory, the changeover takes place granule-by-granule,[39] whereas in membrane transport, it would take place molecule-by-molecule.

Therein lies the story of this experiment. Let me explain. If exocytosis was the mechanism of secretion, then at first, most of the granules whose content was being secreted would have been made prior to the addition of the radioactive amino acid. Over time, more and more of the old, unlabeled granules would be discharged and be replaced by granules that contain radioactive protein. Radioactivity in secretion would reach a maximal value when there were only labeled granules in the cell and only labeled proteins were secreted.

As planned, after several hours of incubating the gland with the radio-active amino acid, Lois added a stimulant of protein secretion. In keeping with the vesicle theory, this was expected to speed the transition from the secretion of unlabeled to labeled protein, accelerating the changeover from the exocytosis of unlabeled to radioactive granules.

It was in light of this expectation that what she found was baffling. Not only didn't this occur, not only didn't the amount of radioactive protein in the samples increase as the vesicle theory predicted, but it *fell* precipitously to almost zero. This was the source of her apprehension and confusion. Had she forgotten to add the sample to the vials or had she made some other error in the transfer? She redid the measurements and obtained the same result. If she had done something wrong, she didn't know what it was.

When she showed me the data, I became very excited. This was no error. She had been carrying out a test of the vesicle theory, and the temporary disappearance of labeled protein certified its dramatic failure. Against its prediction, the stimulant had led to a great increase, not in the proportion of labeled protein in secretion, but to the contrary, a great increase in the proportion of *unlabeled* protein, even as the stimulant had increased protein secretion overall *by over an order of magnitude.*[40]

In the vesicle theory's terms, this was incomprehensible. However, it was readily explained by the equilibrium hypothesis. The increase in the secretion of labeled protein she had been following did not signify the gradual depletion of unlabeled granules as a result of exocytosis. The labeled protein had not come from zymogen granules at all, but from the cytoplasm. What the stimulant did was cause the release of large amounts of unlabeled digestive enzyme from the previously quiescent, older (unlabeled) *zymogen granule* storage pool into the *cytoplasm* where it competed with and overwhelmed the small amount of radioactive protein, whose exit from the cell she had actually been following.

So, it turned out that the cytoplasm, not the exocytosis of granule content, was the source of the secreted protein. Along with the experiments on the disappearing granules, as well as Liebow's experiments on zymogen granule permeability, evidence *against* the vesicle theory and *for* a membrane transport mechanism was growing. More and more, the vesicle theory seemed to be at odds with nature, at odds with the observed properties of secretion from the intact, functioning organ.

At first, vesicle theory adherents found themselves at a loss to explain this observation in terms congenial to their theory. However, some years later, a colleague at UCSF, though neither noting nor citing our experiments, furnished such an explanation.[41] Microvesicles explained the oddity. They not only carried proteins around the cell fusing with various intracellular membranes, but they also fused with the cell membrane, releasing their contents into the duct in a kind of miniexocytosis. The labeled protein did not come from the cytoplasm, but from microvesicles whose content in turn had come from the endoplasmic reticulum. They, not the cytoplasm, were the alternative source. A percentage of these small vesicles bypassed the usual route to the large secretion granules, and instead deposited their contents directly into the duct system.

When the rate of protein secretion was low, secretion came from this bypass system. He called this "constitutive" secretion. As the name suggests, it would take place automatically, being neither dependent upon nor affected by stimulants of secretion. In contrast, he called secretion from granule stores by exocytosis "regulated." It *was* affected by and indeed *only* occurred in response to stimulants.

As applied to Lois' experiment, the stimulant had produced the exocytosis of large numbers of unlabeled secretion granules alongside the ongoing miniexocytosis of microvesicles. But whether writ small or large, whether microvesicles or secretion granules, secretion was due to exocytosis, it was not, repeat *not*, due to the passage of proteins across membranes via the cytoplasm.

This new idea added subtlety (detail and complexity) to the vesicle theory and was seen as significant progress. But it was weak medicine. It was an ineffective antidote, an inadequate explanation for Lois' observation. Continuing the analogy, the vesicle theory remained in the Intensive Care Unit, barely holding on to life. The idea of miniexocytosis posed all sorts of difficulties. To begin with, as with everything about the vesicle theory, there was the panoply of assumptions we have already talked about. One of course had to *assume* that microvesicles existed, that they were what they were said to be, that they moved as proposed and carried the specified substances and only them, and so forth.

Even from a purely anatomical perspective, there was not much to recommend the idea. Microvesicles were not usually found in great abundance adjacent to the duct-facing membrane of the cell, nor did tiny omega figures, no less a profusion of them, dot the landscape. Moreover, such a process would be enormously inefficient. As already explained, it would take hundreds of thousands of tiny vesicles to fill a single zymogen granule. And of course, it would take the same number acting at the cell membrane to secrete the equivalent of the contents of a single zymogen granule.

As a matter of geometry, as spheres become smaller, the ratio of surface area to content becomes increasingly large. Consequently, in regard to vesicles, as they become smaller, more and more membrane is needed to carry less and less content. Microvesicles would approach the physical size limit of such tiny biomembrane-bound objects, with more than half their mass being the enclosing membrane.[42]

The release of the contents of a single zymogen granule by exocytosis would add about 1% to the area of the cell membrane. While the same amount of protein secreted by means of a microvesicle exocytosis would add an order of magnitude more, or about 10%. The area of the duct-facing or apical surface of the cell membrane, where this event would presumably occur, would be almost doubled by the secretion of the equivalent of the contents of a *single* zymogen granule.

And then there was this. What function would such a process serve? Why would it be prominent during periods of secretory inactivity (when it still only accounted for a tiny fraction, as it turns out about 1%, of the total amount of protein being secreted)? Why all this effort, for so little? The whole business seemed absurd.

Yet for believers of the vesicle theory, the concept was greeted warmly. It solved an outstanding problem. And whatever the evidence, it became an established fact of nature. Most importantly, it made it possible to conclude that there was no cytoplasmic pool. Bypass or not, it was all exocytosis, all vesicles, all the time. But this did not end the vesicle theory's travails, not by a long shot. Let's continue.

# chapter nineteen

---

# Irreversible

## The one-way street

> Truthiness: the quality of preferring concepts or facts one wishes to be true, rather than concepts or facts known to be true.
>
> *American Dialect Society, January 2006,*
> *coined by Stephen Colbert on the The*
> *Colbert Report, October 17, 2005*

Whether within or to the outside of the cell, the mechanisms of movement proposed by the vesicle theory occur in one direction and one direction only: forward, towards the ultimate destination of the protein. Having budded from the endoplasmic reticulum, the tiny microvesicles carry their passengers through downstream compartments, fusing, budding, and moving them relentlessly forward to their appointed station. In this, protein movement is irreversible. There is no returning from whence it came. And when secretion occurs, what is released from the cell by exocytosis cannot be retrieved. Once the door is opened, the granule's contents pour out.

This irreversibility is not kinetic. In other words, it is not a matter of circumstance but is built into the vesicle machinery. Add that the events of exocytosis take place *inside* the cell and that stimulants also act there, and we have a simple test of the vesicle theory in the pancreas. *If secretion occurs by means of exocytosis, then preventing the outflow of fluid from the pancreatic duct should have no effect on the rate of protein secretion.* Driven by actions that take place inside the cell, exocytosis should continue unabated, regardless of whether the fluid in the duct flows or does not flow.

The secreted protein would simply accumulate in the standing fluid in direct proportion to the time it sits motionless in the duct. Lois Isenman, by this time a postdoctoral fellow in the laboratory, tested this proposition. Once again, the result was negative (for the vesicle theory) and striking.[43] When she prevented the outflow of fluid from the duct system by elevating the end of the catheter in the pancreatic duct, protein secretion was not only affected, but came to a screeching halt. In the absence of fluid secretion, that is, during the period of stasis, protein secretion had ceased completely!

This was exactly the *opposite* of what the vesicle theory predicted. It had failed yet another test and once again in a dramatic fashion. In the dichotomous world of how molecules move in this microscopic environment, what Lois observed necessarily affirmed the presence of a reversible (equilibrium-based) membrane transport process. It was a concentration gradient between the cytoplasm and duct that produced protein secretion. That gradient had quickly dissipated when fluid secretion was halted, exactly as the membrane transport model predicted.

This result was of a piece with prior tests of the vesicle theory, and I thought would put the final nail in the theory's coffin. To my way of thinking, *this* was the extraordinary evidence that my colleague at Harvard had asked for years earlier. The conclusion seemed unmistakable. Membrane transport processes, not exocytosis, accounted for protein secretion by this cell. But whatever I thought, like everything else, the experiment had little effect on opinion.

In fact, not long after our results were published in *Science* magazine, it published a comment claiming that our interpretation was wrong and that our observations were perfectly compatible with the vesicle theory. The writer said that, despite appearances, neither protein nor fluid secretion had actually stopped! Lois had prevented exit from the end of the pancreatic duct all right, but both exocytosis and fluid secretion had continued unabated. They had simply been rerouted. Instead of passing into the intestines out the end of the pancreatic duct, they exited the duct through paracellular channels, leaky junctions *between* duct cells, and entered the underlying tissue or interstitium.

Again, we were being asked to believe that what was observed was mistaken (the absence of an increase in digestive enzyme content of the duct) and that what the letter's author imagined (both protein and fluid secretion continued without reduction) was real. We were not only being asked to believe what wasn't seen, but also that the ease of egress through tiny shunts between cells was comparable to, the same as passage down the broad duct system. Add that the purpose of digestive enzyme secretion was to digest food in the *intestines*, not the interstitium, and the writer's proposal seemed far-fetched.

Still, the claim required an experimental response. To clarify things, Jenny Ho carried out a series of experiments that paralleled Lois'. But instead of mechanically preventing fluid outflow from the duct system, she decreased the rate of fluid secretion at its source.[44] She inhibited its generation. In this situation, fluid would still pass through and out of the duct system freely, merely at a lower rate. There would be no impetus for diversion.

In this case, if exocytosis accounted for protein secretion, it should have been maintained despite the reduction in flow. If, however, equilibrium processes were responsible, the amount of protein secreted would be decreased in direct proportion to the reduction in flow. The gradient that drives protein secretion would not be eliminated, but diminished in proportion to the reduction. A minor reduction in flow would produce a minor reduction in protein secretion, while a large reduction would produce a large reduction.

This is exactly what was observed. In fact, protein secretion turned out to be *completely dependent* upon fluid flow. No flow, no protein secretion, diminished flow, diminished protein secretion—one in direct proportion to the other. *This was totally contrary to the idea of exocytosis and perfectly compatible with a membrane transport process. Fluid flow was the independent variable that determined the rate of protein secretion.*

This held true even in the presence of a hormone that increased the rate of protein secretion by about an order of magnitude, conditions under which exocytosis should be most active. Though the concentration of protein in the duct system was greatly elevated as a result of the hormone's presence, the response varied in direct proportion to the imposed reduction in flow. So whether it was unstimulated or highly stimulated, protein secretion was dependent on the flow of fluid through the duct system.

As said, this was all just as the equilibrium hypothesis predicted. Still, there was one important point that needed to be resolved. In his letter to *Science*, the writer had cited an interesting paper to bolster his claim. Indeed, its results had probably given him the idea of a shunt in the first place. That paper reported that the pancreas was an especially leaky tissue. Indeed, it was claimed that it was the leakiest tissue discovered to that point.

Unlike other epithelial tissues, such as the skin or gut lining, large molecules easily passed across the gland from blood to duct. It was a very leaky sieve. The authors of the article attributed this to the paracellular channels that the letter writer pointed to. They thought that this made good sense for a couple of reasons. First, in other epithelial tissues, such as the intestine or gall bladder, water and small molecules cross the barrier from blood to the lumen of the organ by moving primarily through leaky paracellular channels. In other words, such channels existed in similar tissues and were responsible for the movement of water and small molecules across it, so why not for large molecules in the pancreas? Second, as a general matter, cell membranes were understood to be far more restrictive barriers than the paracellular shunts, and as already discussed were considered an *absolute* barrier to proteins and, for that matter, any large water-soluble molecule. So, paracellular channels were leaky, while the cell membrane presented an absolute barrier to proteins and other large water-soluble molecules. Hence, they crossed the surface through paracellular channels.

The only problem was that it had not been demonstrated that this was actually the case. Since we knew that proteins passed through the apical membrane of the acinar cell, it occurred to us that the "leak" might be cellular, not paracellular, despite the predisposition to think otherwise. The movement of large molecules might, we thought, occur through, not between cells. A new graduate student, Teri Melese, made finding out what was true her thesis project. She affirmed that the pancreas was indeed a very leaky tissue, and allowed for the passage of large molecules. But she also found that this was due to their passage *across the cells of the gland*, not through paracellular channels![45-47]

So, not only was the fact that the pancreas was leaky to large molecules *not* an argument *for* exocytosis as the letter writer thought, it added new fuel to the fire of disbelief. Cellular membranes permeable to digestive enzymes were the source of leakiness. When all was said and done, Lois Isenman's original observation of the flow dependence of protein secretion by the pancreas *was affirmed*. Everything was as predicted by the equilibrium hypothesis. The vesicle theory had failed yet another critical test. And as a result, once again, the conclusion that membrane transport was responsible for protein secretion was affirmed.

# chapter twenty

---

# Resurrection

## The report of death was premature

Nothing is more sad than the death of an illusion.

**Arthur Koestler**

As important a tool as cell culture has become for the study of biological cells, it is not an all-purpose solution—far from it. Removing a cell from its natural surroundings and anatomical relationships is to lose important aspects of its function. To try to understand how the kidney works in the absence of its complex architecture of organized tubes or to understand how the mind works solely by studying neurons in culture are exercises in futility. You might learn a great deal, but a deep understanding of how the kidney or mind works is not among them. That understanding can only be found where it is articulated, in the organization of each tissue and organ, in the inborn or inherent relationships between like and diverse cells in their natural environment.

Although studying cells embedded in that environment, in whole animals, that is, *in situ*, overcomes this problem, it poses major difficulties of its own. Though time honored and frequently an absolute necessity to gain real understanding, *in situ* research introduces all the unknown and unspecified complexities and confusions of the whole animal and at times makes interpretation challenging at best. There is a third approach to the study of cells, you might say, an intermediate approach that is sometimes used, which has certain advantages over both cell culture and *in situ* research. Unlike cells in culture, it retains the properties of the natural system, while it reduces or eliminates many of the complexities and confusions of research on whole animals.

In using this tactic, small pieces of tissue (or even whole organs, as with the rabbit pancreas) are removed from the animal and studied in beakers, flasks, and other chambers *in vitro*, literally in glass (not uncommonly today in plastic). In this simplified world, the samples retain most of their normal properties. However, this too is no panacea. No matter how much care is taken once a tissue or organ is removed from the animal, it soon ceases to work; it dies. This may only take a few minutes or it may retain its functions for as long as a day, and though different properties often disappear at different rates, its demise is inevitable.

Putting whole organ studies like the rabbit pancreas aside, this method has frequently been used to study the secretion of digestive enzymes by the pancreas. Small pieces of pancreatic tissue are cut from the gland, placed in beakers, and then bathed by physiological salt solutions that are gassed with oxygen. To measure the rate of secretion, the appearance of a digestive enzyme, usually the starch-splitting enzyme amylase, is followed as it is released from the tissue into the incubation medium.

When studied in this way, the rate of secretion falls over time and usually stops altogether in about 2–4 h. This is thought to signify cell death, and hence, in the vesicle theory's terms, the cessation of exocytosis. That secretion coming to a halt connoted death seemed self-evident. But it was not. Indeed, secretion might have stopped completely, while the tissue remained fully alive.

For instance, with an equilibrium-based mechanism of secretion of the sort we have been discussing, cessation might signify nothing more than the elimination of the concentration gradient that drives secretion. Over time, amylase for instance would accumulate in the medium, and as it did, the gradient from the cytoplasm to the medium would decrease and eventually be eliminated, and consequently secretion would stop. So, what appeared to be a dead piece of tissue might simply be quiescent, in an equilibrium state in regard to the secretion of amylase. No amylase gradient, no secretion.

The question then is what causes amylase secretion from isolated pieces of tissue to slow down and to eventually stop? Was it death? In an attempt to answer this question, Jenny Ho carried out another test of the vesicle theory.[48] Once again, the experiments were incredibly simple. The obvious first question was whether despite the cessation of secretion, the tissue still responded to stimulants? Of course, a dead piece of tissue would not respond.

The pieces of tissue were not only responsive, the response was large (an order of magnitude or so), roughly the same as seen when the tissue was fresh (at the beginning of the experiment) several hours earlier. Clearly, the majority of cells were still alive. If not death, what then caused the diminution and eventual cessation of secretion? Was it the result of an equilibrium state between medium and cytoplasm as we thought it might be, or was it something else? For instance, exocytosis might have gradually shut down not due to tissue death, but to the absence of a stimulant.

This was easy to check out. Rather than bathing the pieces of tissue in the same fluid over the several hour course of study, as was usual, Jenny freshened the medium, changing it frequently, thereby removing whatever amylase had accumulated and reestablishing the presumptive concentration gradient. If its accumulation had indeed been the cause of the reduction in secretion, then changing the medium would reinstitute the gradient, and with it, secretion. On the other hand, if exocytosis had shut down, it would have no effect. Fresh medium or not, there would still be no stimulant. It turned out that replacing the bathing medium frequently prevented *any* falloff whatsoever in the rate of amylase secretion. It continued unabated, that is, without reduction over the several-hour course of the study, just as the equilibrium hypothesis predicted.

One uncertainty remained. Perhaps some other, albeit unknown, effect or effects of changing the medium was responsible for the renewal. To check this out, Jenny added a superabundance of a substrate for amylase to the medium. As a result, the enzyme was bound to its substrate busy splitting bonds, and was not free in the medium. This kept the concentration of free (unengaged) amylase low, thereby maintaining the gradient for secretion. As predicted, the falloff in amylase secretion was prevented. It continued without reduction, confirming that its accumulation had indeed been the cause of the reduction in secretion.

This was yet another demonstration that secretion was the result of an equilibrium-based membrane transport process, not exocytosis. The vesicle theory had failed yet another test. To the best of my knowledge, no one has ever attempted to corroborate these extraordinarily simple observations, and as far as I can tell, the erroneous notion that tissue death is responsible for the cessation of secretion remains the common wisdom.

Come to think of it, save for the original report of nonparallel secretion, there has never been any attempt to corroborate any of the experiments we have been discussing. For the most part, they were ignored as development of the vesicle theory continued apace, undeterred, with undiminished intensity. In an ideal world, taking notice of these observations would not be a choice for a scientist, even for vesicle theory adherents, but a necessity, an imperative. But the vesicle theory catechism was impervious, immune to such obligations. They could not be countenanced. They were a hindrance to belief. As Al Gore might say, inconvenient truths had to be disregarded. They were an awkward encumbrance.

Optimists and idealists alike might declare that sooner or later, each of the observations we have discussed, as well as others that we have not, will be given their full and proper due. After all, that is the nature of science. I once believed this with some passion. It was *my* catechism, but nowadays, I am not so sure. I would ask, what in the real world guarantees it? As best I can tell, nothing. The question is whether I am the cynic, or are the cynics those who say that science is inevitably self-rectifying?

## chapter twenty-one

# Pandora's box and the back door

## Now look at what you have done!

> Power corrupts, and science today is the Catholic Church around the start of the 16th century, used to having its own way and dealing with heretics by excommunication, not argument.
>
> **David Gelernter**
> *Commentary, January 2014, p. 25*

Questioning one stricture, one entrenched belief often leads to the questioning of others. When the blinkers are removed, previously inconceivable things can be imagined. And the statement that something is impossible becomes the question, "why not?"

It was in this light that the observation of nonparallel secretion led to the idea that digestive enzymes might on their own, that is, by diffusive mechanisms, cross the apical membrane of the acinar cell. And if this was so, maybe they crossed other membranes in the same way. Moreover, if digestive enzymes could do this, why couldn't other proteins? Indeed, perhaps membrane permeability to proteins was not prohibited after all, but to the contrary was a general property of biological membranes.

And if this was true, then contemporary ideas about membrane and protein structure had to be rethought. Perhaps biological membranes were not cellophane-like wrappers that enabled the cell and its various compartments to hold onto their precious proteins unconditionally. And perhaps proteins were not inflexible polar spheres that were structurally unchanged by their environment. Finally, as it concerned the vesicle theory, if biological membranes were permeable to proteins, the theory's *raison d'être* simply evaporated.

Nonparallel secretion had, as suggested, opened a Pandora's box of speculation. One challenged an understanding of the digestive process that was as old as the discovery of chemical digestion itself. It may be hard to believe, but it was not until the mid-19th century that it was realized that the digestion of food was primarily a chemical event. This came about in great part as the result of research by Claude Bernard of France and Rudolf Heidenhain of Germany, and was one of the most significant consequences of Lavoisier's chemical revolution of the prior century. Along with the discovery of metabolism, it gave rise to modern biochemistry.

Their insight about digestion was based on finding substances in pancreatic secretion that broke food down chemically. They called them "enzymes," loosely translated as "leaveners of life," or "internal ferments." Eventually, it was learned that enzymes were members of a large class of chemical compounds that we know today as proteins. The substances in the pancreas were the *digestive* enzymes, proteins that degrade the various foodstuffs we eat into components that can be absorbed into blood and assimilated by the body. The revelation that digestion was chemical overturned a view, hundreds, if not thousands of years old, that it was mechanical, the result of the exertions of the muscles lining the gastrointestinal tract, most importantly those of the stomach.

As great a discovery as this was, it exposed a previously unknown danger. Since our food and our bodies are made of the same stuff, the substances that digest food are also able to digest *us*, and that of course is a great threat. But somehow we are kept safe. The plain fact is that we manufacture and secrete digestive enzymes and are none the worse for it. Indeed, doing so sustains us.

As traditionally understood, protection is achieved by physically restricting the digestive enzymes to two locations, the pancreas (and other digestive glands), their source, and the gut, their destination. According to this view, the enzymes in the pancreas are isolated in granules, the zymogen granules, where they are kept in an inactive state unable to do harm prior to their secretion into the small intestines. The intestines are defended from them in three ways. First, because digestive enzymes are proteins, they are subject to the same forces of chemical breakdown as proteins in food. So as they break down the proteins we ingest, they digest themselves. Second, special cells lining the intestinal wall secrete a viscous material named mucous that forms a protective layer over the internal surface of the intestines. And finally, the vulnerable cells that line the intestinal tract (the intestinal epithelium) are replaced as rapidly as they are damaged.

This was all nice and tidy. Food was digested and the body protected. But by the middle of the 20th century, an awkward fact came to light. *Digestive enzymes were in blood.* That is right, these destructive substances were not restricted to the pancreas and gut, they were found in blood where they had access to all of the body's internal organs and could digest them just as they did food.

Yet under normal conditions this does not occur. Pancreatic enzymes are present in substantial quantities in the blood plasma of perfectly *healthy* individuals. This raised three questions. First, of course, how do we survive their presence? Second, how did they get there? And finally, what purpose if any do they serve in blood?

It was the third question that sparked an intriguing, but disconcerting idea. It was intriguing because it suggested that an important understanding about chemical digestion might be wrong. And it was disconcerting for the same reason. Any suggestion that this understanding was incorrect would face massive resistance. Even the thought placed us so far outside the mainstream that we would not merely be seen as misguided fools, but as cranks and crackpots. Giving voice to it would further threaten our already endangered efforts to test the vesicle theory and examine the permeability of biological membranes to proteins.

It was one unconventional idea too many. Many of our colleagues would be ill-disposed to it and our work was already fraught. By this time, we were embroiled in far too much controversy. Given our insecure circumstances, the better part of valor would have been to squelch the thought and place it in a drawer of ideas to be prudently forgotten. Evidence that it had merit would not lead to its acceptance, but would instead mobilize resistance, and put our other research in even greater jeopardy. For me personally, it would further damage, perhaps terminally, what little reputation I had left.

Yet, it was impossible to ignore. The concept was relatively easy to check, and it cried out for us to do just that. Having been imagined, it seemed to have a life of its own. To shy away from learning something that might be important just because of what others might think was unsettling. Why, after all, were we doing scientific research? If it was just to comply with the common view, then why not just issue a statement of agreement, or rather a statement of compliance, and close the damn laboratory? Still, curiosity killed the cat, and clearly I did not have nine lives.

Whether foolish or foolhardy, we decided to test the idea, comforting ourselves with the thought (or delusion) that science was about curiosity and that most scientists, committed to learning nature's truth, would eventually accept what we found even if it conflicted with an age-old idea. By this time, experience should have taught me otherwise. But it had not. I held on to this cheerful hope, vainly yet ferociously. They say that once bitten, twice shy, but only if you are sufficiently aware and self-protective.

At any rate, what was the incendiary concept? If digestive enzymes were normal and substantial constituents of blood, not accidental intruders, then what were they doing there? What might their presence signify? The purpose we imagined was potentially important, but it would, as said, in all likelihood ignite new flames of controversy. Indeed, it would not only be controversial, it was combustible, downright incendiary.

Yet it was no more than an analogy to an already established concept. That concept concerned the liver's role in fat digestion. For the fat in our diet to be digested, absorbed, and assimilated, it must first be emul-sified. Detergent-like chemical compounds secreted by the liver into the small intestines called bile salts are responsible for this. Like dish soap, they break large fat globules down into small droplets. Ultimately, tiny microscopic fat-containing particles called micelles are formed, which serve as the immediate source of fat molecules for chemical digestion and absorption.

But the analogy was not about what bile salts did. It was about what happened to them *after* they did it; after fats were absorbed and the bile salts were no longer needed. Bile salts were not excreted as one might expect, passed down and out of the intestinal tract as fecal matter. Instead, they were absorbed into blood and recovered by the liver where they were stored until they were secreted again with the next meal. It turned out that bile salts were used over and over again, meal after meal, with only a small percentage being lost with each meal and needing replacement. This con-servation or reutilization was known as the "enterohepatic circulation."

What we had imagined was an analogous "enteropancreatic circula-tion" of digestive enzymes. Like bile salts, digestive enzymes would be absorbed into blood when digestion was complete, reenter the pancreas and its acinar cell through its blood-facing back door, where it would be stored until called upon for the digestion of the next meal. The case for conservation here was far more compelling than for bile salts. Proteins are by far the most costly biological molecules to manufacture, and their reutilization would save the cell enormous amounts of valuable energy.

Given its potential worth to the body, digestive enzyme conservation was a very attractive idea. There were just two "little" catches. It was impossible in theory and had been shown to be wrong in fact. The tradi-tional understanding was that digestive enzymes secreted into the bowel were *completely broken down* there during digestion and were made anew for each meal.

This conclusion was based on evidence that digestive enzymes disappeared from the gut after digestion was complete. However, this said, there was no proof that their disappearance was due to their destruction. This seemed unnecessary because it was thought that there was no other plausible explanation for their disappearance. First, digestive enzymes were not found in substantial quantities in the large intestine and were not excreted in significant amounts. Second, it was understood that the intestinal wall presented an absolute barrier to their passage. Since they were not excreted and were unable to leave the intestines otherwise, it was concluded that they were destroyed there.

But once again, the phenomenon of nonparallel secretion had changed everything. If the apical membrane of the acinar cell was permeable to digestive enzymes, then perhaps the intestinal wall was also permeable?

Measurements that in the past would have been seen as inane, an attempt to measure what was known not to occur, now were not only reasonable, but called for. Did digestive enzymes cross the intestinal wall to enter blood or was this prohibited? This and the related concept of conservation were matters for observation, for experimental verification, not supposition.

And so we set out to determine whether there was a conservation of digestive enzymes. To confirm the idea, five pieces of evidence were needed. First, needless to say it had to be shown that the gut was indeed permeable to these substances. Second, the digestive enzymes had to enter the acinar cells of the pancreas from the blood side. Third, they had to be stored there in anticipation of the next meal. Fourth, they had to be secreted again with its ingestion. And finally, fifth, the capacity and rates of each of these processes had to be commensurate with that needed to achieve conservation.

Four studies were key. The first was simply to place a radioactively labeled digestive enzyme in the small intestines and look for its appearance in pancreatic secretion. If it appeared, then *per force,* it had been circulated and in the process had passed through the intervening membranes. If it did not appear, then there was no circulation, and we would be saved from another controversy.

Once again, with a combination of exhilaration and fear, Chuck Liebow made the fateful measurements.[49] Within minutes, the labeled digestive enzyme appeared intact in pancreatic secretion, confirming the existence of an enzyme circulation and the permeability of the intestines, as well as the blood-facing (basolateral) and the duct-facing (apical) membranes of the acinar cell to this molecule. The question then was whether this signified a mass conservation of digestive enzymes or whether what Liebow had observed was merely a "trace" phenomenon of no physiological significance.

Making this distinction was the task of the second study. It was the work of Herman Goetze, a postdoctoral fellow from Germany, and his observations were remarkable.[50] He stimulated pancreatic secretion (in fasted animals) and compared the response to the stimulant in three different situations. In the first, or the control, the secreted material was allowed to follow its natural path into the intestines. In the second, it was diverted and collected. And in the third, it was injected into the bloodstream.

He then compared the responses to the stimulant in the three situations. If circulation did not occur, secretion in the three conditions should have been identical. Neither diversion nor injecting the secreted material into blood would be expected to have any effect on the cell's response to the stimulant. On the other hand, if conservation took place, then diversion, removing material from the animal (and from the cycle), would reduce the response. While injecting the secreted material into the animal's bloodstream, bypassing the intestines would restore it. Both were seen— a reduction and a restoration. Indeed, about 90% of what was secreted was conserved. No doubt the most amazing and unexpected result at the time was that injecting digestive enzymes into blood enhanced pancreatic enzyme secretion.

In a third group of experiments, Lois Isenman examined the events at the pancreas.[51,52] She found that digestive enzymes crossed the blood-facing membrane of the acinar cell by means of a reversible diffusion-based mechanism similar to that found at the apical membrane. Not only that, but the process was of substantial magnitude, comparable to that seen at the duct-facing membrane where secretion into the intestines took place.

Next, she observed that what had entered the acinar cells across its basolateral surface was eventually secreted into the duct, entering the intestines in accordance with Liebow's observation of circulation. It turned out that the capacity of this "transpancreatic transport" from blood to duct was enormous. Indeed, it was so great that it could eclipse *maximal* rates of secretion from the gland seen in its absence. For instance, if she allowed the pancreas to accumulate a digestive enzyme across the basolateral surface for a couple of hours before the addition of a stimulant that produced maximal rates of enzyme secretion into the duct, the response was increased, not by a few percent, but *fivefold*! Fivefold greater than a maximal response!

Perhaps the most incredible study of all was the fourth. In it, Kyoko Miyasaka, a postdoctoral fellow from Japan, provided, you might say, a delicious frosting on the cake.[53] She set out to determine the location of amylase in the body after 3h of maximal secretion of digestive enzymes by the pancreas into the intestines (importantly in fasting rats with no food in the gut to digest).

Her findings were extraordinary. What had been released in the intestines was no longer there. It had disappeared. But it had not been destroyed, as tradition would have it. Most of the missing amylase was found, not in the gut, but in *blood*. *Blood plasma had become the second largest amylase pool in the body*, second only to the pancreas and an order of magnitude larger than the intestines. With the cessation of stimulation, things quickly returned to the prior state. The pancreas regained much of what it had secreted, and blood levels fell roughly in proportion. All of this was just as the circulation hypothesis predicted.

These striking observations, as well as others that I have not discussed,[54,55] made it clear that the traditional view that a new compliment of digestive enzymes was manufactured, secreted, and subsequently destroyed with each meal was incorrect. In its wisdom, the body reused these costly substances. After publication of our first paper on the subject, a few groups published experiments that supported our hypothesis, but before long, the idea and experiments came under fire.

Some asserted, without evidence, but also without qualification or doubt that circulation did not occur. Others sought to back their claim with experiments, none of which were failed attempts to reproduce our observations. Different measurements were made under conditions that were not propitious for seeing conservation. But none of this mattered. The affirmative observations were dismissed, and the mere claim that conservation did not occur seemed to be quite enough.

It did not take long before the concept became extremely controversial and that put an end to it. Despite the astonishing observations, we never received, and to the best of my knowledge, no one ever received a research grant to explore the mechanisms and circumstances of conservation. The idea disappeared in short order, leaving only the faintest footprint.

It took 25 years before we published a review outlining the impressive results and explaining why the experiments that had claimed to show that circulation did not occur were misconceived and misinterpreted. Beyond that, we not only explained why conservation did occur, but why it was necessary. The pancreas was incapable of manufacturing a new compliment of enzymes in time for the next meal. It just could not do the job; it could not produce enough. And so, it turned out that conservation was critical for the effective digestion of food in creatures like us.

In any case, it was not long after the initial publication of this research that what I had feared took place. My laboratory was closed. We had not only failed in our efforts to obtain support for our research on the circulation hypothesis but for the lab's work as a whole. In the end, I watched helplessly as fruitful and exciting research came to an ignominious and humiliating end. Not only was I left without the ability to do research, those in my lab, students, fellows, and faculty were left scurrying for positions elsewhere in other fields, and I was unable to help them. A letter of recommendation from me would have sealed their doom.

And so, it all came to a screeching halt, from the studies on the vesicle theory, to those on the mechanisms of membrane transport for proteins, to exploring Pavlov's hypothesis about the regulation of digestion, and of course the circulation hypothesis. In the end, after all the excitement and commitment, after years of work, after all the effort, the entrenched view was even more entrenched. Most disturbing, our remarkable discoveries vanished, sliding, like a sinking ship, beneath the waves, leaving no trace.

*section four*

---

*Reification and attitudes*

# chapter twenty-two

# Expunging heterodoxy
## No you don't!

> It ought to be remembered that there is nothing
> more difficult to take in hand, more perilous to
> conduct, or more uncertain in its success, than to
> take the lead in the introduction of a new order of
> things. Because the innovator has for enemies all
> those who have done well under the old conditions,
> and lukewarm defenders in those who may do well
> under the new. This coolness arises partly from fear
> of the opponents, who have the laws on their side,
> and partly from the incredulity of men, who do not
> readily believe in new things until they have had a
> long experience of them.
>
> **Niccolò Machiavelli**
> *Il Principe, 1505, Chapter 6*

It was so obviously wrongheaded, misconceived, or worse a despicable
attempt to tear down hard-earned knowledge and debase the sterling
accomplishments of brilliant authorities, that it could not be tolerated. To
allow it to continue would be to defame the institution. To give succor to
the heretic would bring suspicion down upon those providing it. Silencing
the pernicious ideas and putting an end to the suspect research was a
necessary and righteous effort. Yes, we are, all of us, open to new ideas
and differences of opinion, but there are limits. This abomination, this
disgrace has gone on for far too long. It has to end!

This was not Rome attacking Galileo, but denunciations I imagined
being hurled at me, a relatively unknown mid-20th century research
scientist, hardly a Galileo, who had the gall to question a modern dogma
of cellular biology. My modest laboratory had been alone in carrying out
serious tests of the vesicle theory and its doors were about to close. The
outrage, the apostasy, was about to come to an end.

The seeds of the downfall had been sown almost as soon as the
questioning began. In the early 1970s, there were rumors that a Nobel
Prize was about to be awarded to the inventors of the vesicle theory. The

gossip turned out to be true, and in 1974, the prize was awarded to one of its founders.[56,57] With the award, the vesicle theory underwent a transformation from a widely held point of view, to an established fact of nature. Its claims were now beyond question, and research skeptical of its propositions was pointless, nonsensical, not to mention a waste of precious time and money. Cell biology needed to devote itself to fleshing out the details of what was now the established vesicle system, not to go on a wild goose chase.

Given this mindset, it would not only have been understandable, but also prudent of me to abandon my reservations about the theory before they were abandoned for me. "Put the skepticism aside and move on" was sound advice offered to me with some regularity by well-intentioned colleagues and friends. But I just couldn't do it. The failed tests remained unexplained, and the weaknesses and ambiguities of the evidence for the theory had not evaporated overnight with the awarding of the prize, however prestigious. Not only that, but by this time, I had come to the conclusion that the vesicle theory was fundamentally misconceived.

And so, however imprudent, however poor the prospects and unpromising the journey, I decided to press on. For this to be possible, I had to find some way to publicly—that is, to a large scientific audience—rebut the proclamation that the vesicle theory was settled science. I had to mount a counterattack, to make the basis for my doubts more widely known. Towards this end, I decided to write an article for the prestigious and broadly disseminated *Science* magazine as a riposte to the encomium it had just published on the research of one of the recipients of the Prize.[58]

Though I was worried that this might be viewed as discourteous, my intentions were merely to keep the door open, to allow the theory to be challenged despite the acclaim. I don't know if my arguments convinced many people, but in the wake of the article's publication and despite the continued presence of strong headwinds in the form of critical editors and grant reviewers, I was able to continue the research, publish the results in first-rate journals, and even expand the effort modestly for almost a decade. At times, the ability to persist seemed almost miraculous, with the miracle not being of my making, but nature's. *Its* properties propelled our work. They allowed it to continue despite the long and deepening odds, the continuing disparagement, and the unending attempts to put an end to it all.

Still, as they say, all good things must come to an end, and the end was nigh. Though over the years experiment after experiment had strengthened our argument against the vesicle theory and the scope of our research had expanded, the guillotine was poised, and the blade was about to fall. Though I had harbored the fear of being shut down for a long time, a worry made worse by the circulation hypothesis, I was clueless about our imminent demise.

Actually, things seemed to be looking up. We were better situated than ever, better able to defend our thinking and research. Yet, it was all about to come crashing down. And this was not due to our failure, but the opposite—our strengthened position. The work had become increasingly dangerous. The sedition could no longer be tolerated.

For the responsible institutions and individuals, closing the laboratory was simple in a mechanical sense. Remove its contents and change the locks. However, there were problems of appearance and propriety that could not be ignored. Closure had to be carried out in a way that would leave no doubt as to the unfailing commitment of those involved to free and open inquiry. They could not be seen as being vindictive or close-minded.

Most importantly, their decision could not look political. This was not Rome, but neither was it the 20th century's enforcer of scientific orthodoxy, the Soviet Union. I was in the enlightened West, where free thought flourished. The decision had to at least have the appearance of having been based on a rigorous and unbiased assessment of the validity of the ideas and evidence.

The contrivance that was used to achieve this goal, to provide the necessary cover and pretense, was "peer review." In peer review, experts in a particular field evaluate the efforts of their colleagues, their peers. Are their ideas and methods creditable? Do they significantly advance scientific knowledge? Is the work praiseworthy? Should it and its originators be supported, promoted, and even honored? Peer review provided an impartial means of answering such questions and of evaluating individuals and their work.

It also made it possible for the responsible persons, administrators, bureaucrats, and editors to say with comfort and authority that their decisions were based on the best available *scientific* opinion. They could say that they had no desire to silence anyone or any thought. They were simply reflecting the will of the relevant scientific community. Their decisions were scientific. They were based on reason, not politics.

In science, three agencies are responsible for these evaluations. First and foremost, there are the scientific committees that review research proposals. They are usually constituted by bureaucrats in Washington, DC and are comprised of scientists from universities and other research institutions around the country and the world. These committees, not the paper-pushing bureaucrats, determine which research projects deserve funding and which do not.

Next, there are the scientific journals, with their reviewers and editorial boards. They are responsible for determining what is and what is not worthy of being published. And finally, there is the university with its administrators assigning precious laboratory space and resources, and its faculty committees evaluating personnel for promotion. Membership on each of these peer review bodies—research, publication, and promotion—is in great part determined by word of mouth, as scientists at various universities and institutes are called upon to evaluate the work of others in their area of expertise.

What is most important to understand about peer review committees is the basis for their decisions. What research is valuable and what ideas are credible, and which are not is determined by *majority opinion*. The most significant, indeed the most remarkable thing about peer review, is that for all practical purposes what is legitimate and what is specious, *what is true and what is false is up for a vote*. This democratic process governs not only science, but also what it understands to be true. If an idea or evidence does not pass popular muster, it is deemed incorrect, untrue, whatever the actual facts. You might say, how horrible, how unscientific, how irrational! But however unscientific, however irrational, this is the way it is.

Yet despite such a conspicuous and lamentable flaw, peer review is invariably lauded. The reason is simple. However execrable, it is the best we can do. How else, it would be asked, can we evaluate scientific efforts? What would replace it? Yes, it is unfortunate that at times its votes give malign and iniquitous results, but this is a necessary inconvenience. There is no better way to evaluate scientific work.

Would we be better served, if the decisions were left to individual powerbrokers, to princes or despots? Or should they be based on how well positioned the individual scientist is in relation to the controlling authority of the state or the academy? Would it be better if it was a matter of class, not talent or accomplishment?

It can be said, and with a great deal of justification, that the introduction of peer review represented a great advance for science. It was the end result of much effort and struggle to disentangle science from social prejudice and favoritism. It allowed science to move beyond far less enlightened centuries-old circumstances. Today, scientific decisions are made by groups of *experts*, not despots, dukes, or kings. They are not determined by one's position in a social or political hierarchy, but by merit.

But whatever its promise, however noble the goal, has peer review really ridded science of biased decision making, or has it merely changed the basis of the bias? To think that its decisions are the result of a purely logical exercise, devoid of bias and political considerations is to live in Cloud Cuckoo Land. When you get down to it, looked at with detachment, the process gives voice to the expression of all kinds of biases, scientific, political, even hierarchical. A decision about a particular line of research may rest not on scientific facts at all, but on whose opinion matters, whether it is the opinion of a powerful individual or a group of like-minded experts.

The overarching problem is that it is in the nature of experts to have strong points of view, particularly about subjects in their sphere of expertise. Most often, these are not easygoing, lightly held viewpoints, but fervent and committed ones. Worse, experts frequently have a stake, even a formidable personal stake in the outcome of an evaluation however well hidden in the language of scientific probity.

Certainly, if what a particular individual has done or proposes doing *runs counter* to the perspective or interests of an expert reviewer, no less a group of them, peers however collegial are not disinterested parties. The expectation that they will be broad-minded and open to different ideas, as opposed to being prejudiced, judgmental, and narrow-minded is a happy, but unconvincing thought. On the other hand, if what the particular individual has done or proposes doing supports or compliments the reviewer's efforts, desires, and dreams, if it is agreeable to his or her work and ideas, are we to think that this will not have a salutary effect on the evaluation? Needless to say, agreeable evidence and proposals are more likely to receive an approving, if not an admiring evaluation, than work and ideas that conflict with the expert's views and position.

This does not make peer review demonic. The differing attitudes of reviewers towards the comfortable and agreeable and the uncomfortable and disagreeable are no more than human nature. With expertise come attitudes, good and bad, justified and unjustified, but either way, it comes with strongly held convictions. Whether an evaluation is complimentary or censorious, the expert comes fully loaded for bear. And so, although scientific observation has a basis in reality independent of the attitudes of its practitioners, in peer review, majority opinion, with its accompanying self-serving and political baggage, not reason, is the arbiter of truth.

Still, the system usually works well enough. It provides what I call a balance of terror. If you speak admiringly of me and my efforts, I will return the favor and all will be well, research will be done, publications produced, promotions granted, funds will flow, and progress will be made. But if you question me, I will question you, and we will both lose. Yes, I will point out weaknesses in your efforts, as you will in mine. Not to do so would be unprincipled and unethical, but I will also be sure to note that these are minor problems that you, as a skilled researcher, will doubtless correct.

This process is reasonably effective for those who share the same general outlook or approach, for those in the business of Kuhn's normal science. In normal science, the line of attack and methods may vary from lab to lab, but the overarching goal is the same. Everyone is seeking evidence to fill in the holes in the various models and theories of the paradigm. Research is designed to bolster its claims, not question them.

Where peer review fails, and fails abysmally, where it is anything but fair, is for the small number of scientists who demur, who disagree with or take exception to the common view. For them, for those who question the paradigm, for those who are not engaged in normal science, peer review is invariably and patently unfair. Not only doesn't it work to secure their interests, it is a powerful weapon that can be and usually is marshaled against them.

Even when such dissidents serve on review bodies, they represent a small minority, most often a minority of one, and have little real influence, no meaningful say, and certainly no leverage about what judgments will take place. This said, most often they are excluded from these deliberations altogether. For myself, despite being on the faculty of two of the world's most prestigious academic institutions, I was never asked to evaluate the work of those with whom I disagreed. I was never asked to serve on government committees that determined the funding of their research or to comment on their promotion or the awarding of academic honors. Nor did journals ever ask me to review their research work. Never, and I mean never, in some 50 years was I asked to evaluate the work of supporters and developers of the vesicle theory.

On the other hand, their meat hooks were all over my research and my career and that of those who worked with me. If I complained that one or another person was biased and should not be involved in a particular evaluation, I was told that it was not appropriate to exclude them just because we disagreed. This was science after all, and the purpose of peer review was to provide a fair evaluation of the research and researcher, and this could only be done if a range of opinion was sampled. In fact, if anything, it was more important to secure critical opinion for research that questioned the common view.

Not only that but requests to exclude individuals from the evaluation were seen as improper, unprofessional, and even unethical. It reflected an attempt to bias the proceedings by shutting out those with whom you disagreed. On the few occasions where my complaints about reviewer bias were honored, it was usually in the breach. Others, who held the same negative view of our effort, more often than not their collaborators, simply took their place.

From time to time, the decision makers complained in mock or real exasperation that they were having a very hard time finding truly unbiased individuals to review our work. And no doubt this was often true, at least for the world of experts on the particular subject. The sad fact was that we were without allies, and those with an open mind faced danger by publicly expressing their doubts about the paradigm by reviewing our work positively. It was safer to keep such feelings private, not to get involved.

For the iconoclast, for the odd man or woman out, in science as in life otherwise, there is no level-playing field, and peer review is certainly not a means of leveling it. To think that modern science is somehow exempt from the biases and prejudices of human nature, is either naïve or dishonest. Beyond this, if the goal is to get rid of fallacious or dangerous ideas, the unfair or unsavory may not only seem necessary; it may seem principled and even virtuous. And so on to the closing of my laboratory.

# chapter twenty-three

# The visit

## The unwelcome guests

It is not easy to discover from what cause the acrimony of a scholiast can naturally proceed. The subjects to be discussed by him are of very small importance; they involve neither property nor liberty; nor favour the interest of sect or party. The various readings of copies, and different interpretations of a passage, seem to be questions that might exercise the wit, without engaging the passions.

But whether it be, that small things make mean men proud, and vanity catches small occasions; or that all contrariety of opinion, even in those that can defend it no longer, makes proud men angry; there is often found in commentaries a spontaneous strain of invective and contempt, more eager and venomous than is vented by the most furious controvertist in politicks against those whom he is hired to defame.

*Preface to Samuel Johnson's Edition of Shakespeare's Plays, Page lvii, Printed for J. and R. Tonson, H. Woodfall, J. Rivington, etc., London, 1765*

Though we had become an isolated island, we had our work, and it encouraged us. The wonders of nature were comforting. They gave us hope, even joy, day in and day out. But demoralization came inexorably and inevitably. With the closure of the laboratory, we found ourselves, quite literally, out on the street. That we were victims of what I thought was other people's bad behavior offered cold comfort, and I still find it painful to tell the story even from this remove of time. Although the shame was theirs, what happened was not merely embarrassing and humiliating, I had been stigmatized, and my ability to do research seemed to have come to an abrupt and final end, whatever my desires and interests.

How did this sad state of affairs come to pass? It was a typical day at the University of California, San Francisco (UCSF). Parnassus Heights was cold, windy, and enveloped in fog. We were rehearsing our presentation for a site visit from the National Institutes of Health (NIH) the next day. We had become increasingly productive during the past few years. Each experiment seemed to open new doors, and our challenge to the standard view had become stronger and stronger. One result of this progress was that the small research grant that had supported the laboratory for years had become insufficient. There was not enough money to maintain the current level of research activity, no less accommodate the significant expansion that I felt was urgently needed.

With that money alone, we were not going to be able to carry out the research required to achieve any of our major objectives. We were not in a position to establish the details of the mechanisms involved in the passage of digestive enzymes across membranes, no less develop a full understanding of the regulation of digestive reactions, or characterize all the features of the enteropancreatic circulation. It was all beyond our capacity.

Worse yet, renewal of our small grant was at hand, and the likelihood of success was not merely uncertain, it was doubtful. After all, a decade had passed, and we had not changed minds, most importantly, the minds of the experts and opinion makers. And, as said, the paradigm not only remained in place, it had become more strongly embedded in the scientific psyche. If our criticism had truly been justified, if the evidence we had accumulated was in fact dispositive, then why, it seemed natural to ask, hadn't attitudes changed?

It was in this foreboding climate that I heard of a new request for applications from the NIH. A small number of grants were to be awarded in fields that included the study of pancreatic secretion. They were called Program Project Grants. Each grant was intended to provide funding not just for one project, but for a range of projects with a common theme. This was just what we needed. It would not only permit the continuation of our work, it would provide for its much-needed expansion.

Perhaps, one of these grants could be awarded to a research program like ours that sought to challenge, rather than reinforce, the standard view. With the encouragement of administrators at the NIH, we decided to apply for one of the new grants rather than trying to renew our current, inadequate, and soon-to-expire grant. The proposal we produced was wide-ranging and included research projects on many of the themes I have talked about.

In any event, the day of reckoning, the site visit, was upon us. A team of visiting scientists would evaluate what we had proposed the next day. We would soon find out whether the optimism we felt was warranted. Much to our chagrin, it was not—far from it. What we did not appreciate was that this attempt to expand the laboratory's efforts to accommodate its successes had sealed its fate. The application shone a bright light on the strong bias against our point of view and its challenge to the standard opinion. It provided an opportunity for administrators and scientists alike to pull the plug, to end the work, and to sanctify the abnegation.

Nonetheless, in a blissful state of denial, we hoped for an unbiased assessment of our research proposal. Unbeknownst to us, behind each administrative assurance of support, and there were many, was a hidden commitment to prevent the grant from being approved and funded. Our effort was simply not going to be successful, and there were plenty of clues if one paid attention.

The first clue—actually, it was more like a frontal assault than a clue—occurred when a colleague at UCSF, another member of my department, informed me that he was going to apply for the same research grant. The likelihood of two large grants on the same subject being funded at the same institution, especially when there was only enough funding for a few nationally, was, to say the least, very unlikely. As if this were not enough, not long after he informed me of his intentions, on the day before Christmas, I received a call from the chairman of my department, executing a powerful second attack.

Holiday or not, he was assigning me a major new teaching responsibility on a subject that I had not taught for 10 years, in a course for some 150 medical students that was to begin the next week, on January 1st. At this level of instruction, it was common for professors to be given time, usually six months or so, to prepare for a new assignment. I had been given a week. I could not simultaneously produce a satisfactory grant application for the NIH deadline at the end of January and carry out my new teaching responsibilities, even in a slipshod fashion.

But if I did not submit the application, funding for my laboratory would end in a few months. I was stuck, as they say, between a rock and a hard place. Given what I saw as the gravity of the situation, I decided to ask the dean for his advice. He was reassuring. Don't worry, he said, just work on the grant. He would deal with the department chair about the teaching assignment. I took his comforting advice to heart and finished the large and daunting task on time.

Now here we were, a few months down the road, and our efforts were about to be judged. The run-through of our presentation had gone well and I felt good about our chances. The proposal was interesting, filled with new ideas, and powerful evidence. But outside of our small and isolated world, approving such a substantial grant for our work was out of the question. It would not be seen as progress, as being of significant scientific value, but as an enormous threat. It was one thing for the NIH to fund a small grant supporting research that questioned a widely held point of view, but quite another for it and, for that matter, the university to support a major expansion of the effort, and that was what we were asking.

Awarding the grant would give the work prominence, but more importantly, it would signify that the NIH and the University were bestowing their collective imprimatur on the effort. It would be saying that our viewpoint was valid. It was one thing to be open-minded, but to give the transgression such a stamp of approval was, as we were soon to find out, unthinkable.

The dénouement came at the site visit. I had the first inkling of trouble when the committee's members were introduced. Against my protests, a collaborator of the vesicle group, a senior scientist to whom other members of the committee might well defer, was on the committee. There was no way that he would be convinced of the validity of our work or the reasonableness of our ideas even if they had come directly from the hand of the deity.

Two deans were present, and that was reassuring. I thought their support might help. The presentations of the various projects went well, and we were able to answer the committee's questions. But the blade had yet to fall. Just before the meeting was adjourned, the chairman of the visiting committee asked the fateful question: Had the deans made plans to provide additional research space for the project?

They seemed taken aback by the question. They said that while they supported the application, they could not commit space in advance of the grant being approved. Only then would they consider such a request. I had not thought seriously about the need for more space and had never had more than a fleeting conversation with either dean, or, for that matter, anyone about it. It was the last thing on my mind. Of course, I would seek more space if we obtained the grant, but with or without it, we would manage. Whatever soothing words they said to the committee to the contrary, their unwillingness to commit resources was viewed accurately, I think, as a lack of administrative support.

And so, the proposal was not funded and the insufferable research was to come to an end. The thorn would be removed. And most importantly, neither the university nor the NIH could be held blameworthy. They were just doing their jobs, following the advice of the peer review committee. The next summer, on returning from a pleasant vacation with my family, my laboratory was gone, dismantled, my files stored God knows where, and my personal belongings placed in a tiny, windowless closet that was to serve as my office. I was without funds for this or any research, my laboratory was gone, and despite having tenure, my position was threatened.

My students, fellows, and colleagues were sent into the wilderness. They had to seek other positions. They could not pursue their current line of research, no matter how promising and whatever their desire. There was no possibility of funding, anywhere, under any condition. Moreover, due to no fault of their own, their connection to me and to the work had diminished their job prospects. As those who had come before them, they either had to disavow the apostasy (some did) or silently move on to another field.

Though I doubt that any of those involved in determining the fate of my laboratory and its research harbored any guilt, I thought that the main actors in the play had acted in disgraceful and dishonorable ways. Perhaps some thought that what had happened was unfortunate, even sad, but necessary nonetheless. The time was long overdue for the unacceptable research to end.

Doubtless there are many in science, in the academy, not to mention the world outside, who have had far more ghastly experiences for having had the temerity to disagree with the accepted view. As for science, historians tell us of the sufferings of many Galileos, but such talk is invariably couched in the past tense. The implication is that in our own time things are different. Today, we are open and fair.

We pat ourselves on the back and credit our enlightened and open minds for this benevolent state of affairs, proudly exemplified by the jury-like actions of peer review committees. Dissenters are after all no longer incarcerated or put to death (though there have been legal proceedings). Nevertheless, despite all the self-congratulation, science still has a great deal of difficulty coping with iconoclastic views. To question a common scientific belief remains a perilous business.

As things transpired, the other application from UCSF *was* successful. The work it proposed contained not the faintest hint that the vesicle theory was other than an established fact of nature. Anyway, at the end of the day, I faced the unenviable task of reimagining my scientific life.

# chapter twenty-four

# Rejuvenation
## What now?

> Hope springs eternal in the human breast:
> Man never is, but always to be blessed:
> The soul, uneasy and confined from home,
> Rests and expatiates in a life to come.

**Alexander Pope**
*An Essay on Man, 1734*

I decided to take a long-overdue sabbatical across the Bay at Berkeley to think about my future. Restoring my laboratory at the University of California, San Francisco (UCSF) seemed impossible. Everything was gone, not only equipment, files, and so forth, but people, students, technicians, and fellows. More importantly, I had no supporters, no less advocates either at the university or the National Institutes of Health (NIH).

I was defrocked: a laboratory scientist without a laboratory. My effort to test the vesicle theory and characterize the mechanisms of membrane transport for proteins was over, and so, it seemed, was my research career. Despite what I believed to be substantial accomplishments, and however instructive our discoveries had in fact been, in the social and political world that determines what science is worthy and what is not, I had failed miserably. My efforts had been judged unworthy.

As a result, what I loved had been taken away from me. The question was whether I could somehow turn adversity to advantage. After all, I had been freed from the shackles of controversy and the responsibility of running a laboratory. Perhaps I could develop my other, nonlaboratory scientific interests, or find a different research path. One day, amidst my pondering, exploring, and agonizing, I noticed a seminar at Berkeley to be given by a physicist, David Attwood, from the Lawrence Berkeley Laboratory (then LBL, but now LBNL to signify that the laboratory is a "national," not just a Berkeley, resource). His talk was about a potentially groundbreaking new microscope, an X-ray microscope—Superman writ small. Instead of using light from the visible spectrum like ordinary microscopes or electrons like the electron microscope, it used X-rays in a particular range of wavelengths called "soft X-rays" as the light source.

The idea was not new, but it had only recently become possible to build an X-ray microscope that could live up to its theoretical potential.

This was made possible by two technical advances. Unlike the hard X-rays of medical X-ray machines, it was possible to focus beams of soft X-rays to a fine point using minute diffractive lenses (Fresnel lenses) fabricated by the methods of deposition and lithography that had recently been developed by the semiconductor industry. Second, to focus the light effectively, the X-ray beam had to be coherent. This meant that the light had to be of the same wavelength and in phase. A certain type of particle accelerator, a synchrotron, originally developed to study the intimate structure of the atom, was well suited to this task.

In theory, resolution close to that of the electron microscope could be achieved with soft X-rays. But it was not resolution alone that was the reason for interest in the new microscope. It offered the promise of a high-resolution microscopy that could *overcome some of the intractable limitations of electron microscopy*. Biological samples could be viewed in air, not under vacuum, and without the need for sample preparation. High resolution could be achieved with samples in their natural state, whole and suspended in physiological solutions. Most exciting was the possibility of looking at things at high resolution *as they happened*. In addition, the method opened the door to a microscopic X-ray spectroscopy that could estimate various chemical properties of the tiny samples. Of particular interest to me was their protein content.

Though the new microscope seemed to have much promise, there was also a great deal of skepticism about the claims being made for it. At Attwood's seminar, a member of the biophysics faculty at Berkeley articulated this skepticism with an angry certainty. He said that before coming to the seminar, he had made a calculation, and the energy required to image biological objects with soft X-ray beams would vaporize them. For all Attwood's hype, the microscope was useless for biology. However condescending and firm in his conviction, and he was both, the questioner did not reveal his calculation, as it turned out for good reason.

He was wrong. The device could work without destroying biological samples. Indeed, the mass loss (destruction) produced by the energy of the beam was a far greater problem for electron microscopy. The question was not whether the sample could be preserved, but whether the method could be made practical, could high-quality lenses be made, could the coherence be produced, could samples of various kinds be mounted in the device, and could detectors be developed to collect and analyze the information at the promised resolution?

As I sat there listening, I began to imagine exciting measurements on zymogen granules with the new microscope, measurements that had not been possible before. With much regret, I had concluded that despite clear evidence that the zymogen granule membrane was permeable to the proteins it stored, as long as one could *imagine* granules fracturing, breaking apart, the claim of artifact would be the accepted explanation for release whenever it occurred. Whatever had been proven to the contrary was irrelevant. The mere possibility of an artifact, of lysis, of what could be dreamt up was sufficient.

But if one could actually *see* intact granules releasing their contents, could the truth still be denied? They say that "seeing is believing," and if this is true for anyone, it certainly is for microscopists who spend their lives *looking* at things. The X-ray microscope seemed to offer the electrifying possibility of actually seeing granules release their protein contents. With this exhilarating thought in mind, I approached Attwood about the possibility of using the new microscope.

Unfortunately, it turned out that its construction was years away. He had been talking about something that did not yet exist. The Department of Energy was about to build a new synchrotron light source at LBL on the skeleton of the old Cyclotron, where the atom was first smashed. It was Attwood's plan to build an X-ray microscope at this new Advanced Light Source. Though the light source, no less the microscope was years in the future, he pointed out that across the country, all the way out on Long Island, at Brookhaven National Laboratory, a group of physicists led by Janos Kirz of SUNY, Stony Brook, was developing an X-ray microscope, and it was about to become operational. He said that his group at LBL was involved in their project, and that he would be pleased if I joined them and carried out the experiments I had imagined at Brookhaven.

And so, I found a new home and a new laboratory on the other side of the bay at LBL, as well as on the other side of the country at Brookhaven among friendly and extraordinarily bright physicists. With some modest funding, and a new graduate student, Kaarin Goncz, who was studying for her Ph.D. in biophysics at Berkeley, the project got underway. While the physicists and engineers at LBL were feverishly designing and assembling the new light source, we began our experiments at Brookhaven. We would soon see what happened to zymogen granules as they released their protein contents or at least that was our hope.

Charles Liebow had shown that the flow of fluid past populations of granules produced the continuous release of their contents. By removing granule contents this way, we thought that we should be able *to see* the effect of release on the stability of the granules. Using methods of nanotechnology, Kaarin developed a tiny viewing/perfusion chamber in which we could observe individual granules as fluid flowed past them.[59]

Did perfusion lead to granule lysis, to the destruction and disappearance of the granules, or did they remain intact as they released their

*Figure 24.1* Contemporaries of Heidenhain, Wilhelm Kuhne, professor at Heidelberg (photo) and Arthur Sheridan Lea, professor of physiology at Cambridge, reported that granules shrink with secretion.

contents? Either way, we would know the truth about release, about lysis. After much effort and struggle with the complex and demanding machine, she succeeded. As granules lost their contents, *they shrunk!*[60]

*That's right they shrunk.* Though apparently unknown to most devotees of the vesicle theory, even experts, knowledge that granules can shrink was not new. It was first reported in the 19th century by contemporaries of Heidenhain, by Kuhne in Germany (the man who coined the term "enzyme") and by Lea in Great Britain, in, believe it or not, the rabbit pancreas.[61] More than a century later, Thomas Ermak in my laboratory carried out a remarkable series of studies using the electron microscope that showed that granule size decreased with the increases in digestive enzyme secretion produced by feeding,[62] increased as the pancreas began manufacturing digestive enzymes *in utero*,[63] and dramatically decreased with the onset of secretion after birth (Figures 24.1–24.3).[64]

Kaarin's observations confirmed Ermak's and the connection between granule shrinkage and the secretion of digestive enzymes. But more important to us at the time, there was no lysis! Granules did not break apart or disappear. She was even able to reverse the process. By adding a digestive enzyme to the perfusion medium, granules increased in both size and protein content.

What Liebow had shown in his kinetic experiments could now be witnessed happening. It was possible to actually *see* intact granules losing their contents, and lysis was nowhere in sight. So as it happened, almost 150 years after Heidenhain's original observations of disappearing

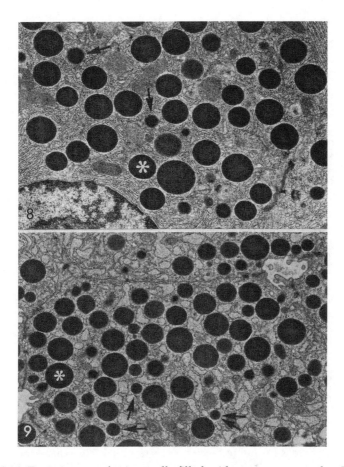

*Figure 24.2* Two pictures of acinar cells filled with zymogen granules. The top image is from fasted rats and the bottom ones from fed rats. Do you see the difference between the two? The asterisks are intended to help. They mark granules of roughly the same size in the two images. It is the largest granule in the fed sample and a less than average-sized granule in the fasted sample. Arrow in top image points to granule not sectioned through the equator and hence larger than shown, whereas arrows in the bottom image show small granules sectioned through the equator. Granules shrink as enzyme is secreted (feeding). (From Ermak, T.H. and S.S. Rothman. Zymogen granules decrease in size in response to feeding. *Cell Tiss. Res.*, 214:51—66, 1981.)

zymogen granules, his explanation for their disappearance, as well as its modern incarnation, exocytosis, was shown to be incorrect. Granules did not leave the cell *in toto*. Though they had seemed to disappear in the wake of high levels of enzyme secretion in Heidenhain's light microscope, its limited resolution had made it impossible for him to distinguish

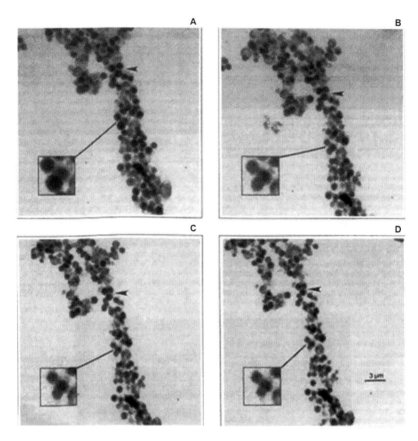

*Figure 24.3* Zymogen granules seen in the X-ray microscope shrinking and losing protein as water flows past them (in order of time, figures A–D). (From Goncz, K.K. and S.S. Rothman. Protein flux across the membrane of single secretion granules. *Biochim. Biophys. Acta*, 1109:7–16, 1992.)

between decreases in granule number or size. All he really knew was that the portion of the cell they occupied had decreased.

Yet, as with everything else we have discussed, Kaarin's observations changed nothing. What took place was there for all to see, but it mattered not a whit. Acceptance was still not in the wind, and work filling in the details of the putative vesicle mechanism continued without the slightest hesitation or second thought. In fact, in the years following her observations, two new Nobel Prizes were awarded for discoveries related to the vesicle theory. The theory was so firmly ensconced by that time negative evidence, whatever it might be, was irrelevant. No matter how many tests the vesicle theory failed, the mechanisms it proposed were seen as facts of nature. And the idea that secretion involved the transport of proteins across membranes was understood to be false, not as a matter of evidence, but by declaration.

Kaarin's project had been done quietly, under the radar (or X-rays, if you will). But as she started to prepare her thesis and publish her results, and as the Advanced Light Source was approaching completion, other biologists at LBL and on the Berkeley campus became aware, not only of her work, but of my presence, and that was not a good thing. It did not take long before I became an object of attack. A controversial scientist, a nettlesome outsider from San Francisco who should not be involved in, no less lead the effort to develop biological research for the microscope at the new light source. That had to be stopped.

And so, it was not long before my pleasant and welcoming sojourn at LBL ended. And in a final, breathtaking irony, the managers at LBL decided, in their wisdom, to have the effort to develop the microscope for biological use led by the scientist who had dismissed the idea at Attwood's seminar, and whose attitude about its potential utility remained ignorantly, arrogantly, and unflinchingly negative.

# chapter twenty-five

# Pavlov and the zymogen granule
## Could it really be so simple?

> Never give in. Never give in. Never, never, never—
> in nothing, great or small, large or petty—never give
> in, except to convictions of honour and good sense.
> Never yield to force. Never yield to the apparently
> overwhelming might of the enemy.
>
> **Winston Churchill**
> *Harrow School speech, October 29, 1941*

I have no idea what Pavlov thought about zymogen granules, or even if
he was aware of their existence. But whatever he thought or knew, they
appear to be key to the regulation of digestive reactions.

Hsien Tseng, a visiting scientist from China, Jim Grendell, then
a fellow in medicine, and Joel Adelson, a young biochemist who had
accompanied me to the University of California, San Francisco (UCSF)
from Harvard, variously discovered that particular peptide hormones,
not only cholecystokinin (CCK), but insulin, glucagon, and a new peptide
isolated from the duodenum named chymodenin, led to the preferential
secretion of specific enzymes.[65–67] Importantly, it was also discovered that
*end products* of digestion themselves, such as glucose and various amino
acids, also had a selective effect on secretion.[68] Small amounts injected
into the artery that immediately feeds the pancreas favored the secretion
of certain enzymes.[69]

Unlike most biochemical reactions, the reactions of digestion are not
controlled by negative feedback, where reaction products limit their own
production, but by positive feedback. In positive feedback, the rate of the
reaction in increased. For digestion, this was due to an increase in the
rate of *secretion* of the relevant enzyme. Unlike positive feedback in other
chemical reactions that often results in overheating and disaster, even
fires or explosions, nature prevented such an occurrence here. There was
an off switch, a way to turn the process off that prevented not overheating,
but overeating. Of course, I am referring to the sense of fullness or satiety
that ends a meal, which causes us to pull back from the table.

In any event, the realization that end products of digestion can act directly on the acinar cell led Jim Grendell and Klaus Niederau, a post-doctoral fellow from Germany, to wonder about a fascinating possibility. What if the *zymogen granule* itself was involved in this process? What if digestive reactions were regulated by the differential release of enzymes from *the granule*?

This was an exciting idea, and easy to test. Just collect granules and suspend them in a fluid that contains a particular end product and look for the selective release of digestive enzymes from the granules into the medium. When this was done, selective release was found. Particular end products produced the release of certain enzymes from the granule, leaving others in place, unperturbed.[70]

How exciting! It seemed that the digestion of a meal was at least in part regulated in this elegant and simple way. This was consistent with observations made many years earlier, in which the pH of the medium in which granules were suspended had a differential effect on the release of their contents. Release did not occur en masse at a given pH, as one would expect if lysis were responsible, instead the pattern of release was unique to each enzyme.[71,72]

At any rate, Pavlov's theory found a welcoming home in this tiny object. This presented a wonderful opportunity. At least in theory we could study the whole range of molecules that we thought might play a role in the regulation of digestion by examining their effect on the lowly zymogen granule in this relatively simple way. Despite our desire, this aspiration was not to be realized, and these studies were never carried out. With the closing of my laboratory and the whole enterprise under a dark cloud, as with everything else, studies on the amazing zymogen granule came to an end with a whimper, not a bang.

Let me end this sad story with one last disappointment related to the study of the zymogen granule. Kaarin Goncz' experiments showing the release of protein from single granules offered another great opportunity. The ability to measure the protein concentration of individual granules and follow its release made it possible to establish the mechanisms of digestive enzyme transport across granule membrane with kinetic and thermodynamic precision.[73] But this, too, was not to be, as our research at Lawrence Berkeley Laboratory came, as said, to an inglorious and shabby conclusion.

This said, the matter did not end there. The vesicle theory and its insistence that membranes were impermeable to digestive enzymes (and for that matter all proteins), however widely believed, remained weak. Indeed, the theory had problems that went beyond experimental observations. There were significant and in some cases limiting theoretical issues. Let's take a brief detour to see what I mean.

# chapter twenty-six

# The impossibility of it all
## On the coherence of theory

> I learned very early the difference between knowing
> the name of something and knowing something.
>
> **Richard Feynman**

Throughout this book, I have focused discussion of the vesicle theory on experimental observations. Is the theory compatible with the properties of nature as they are seen in laboratory experiments? But there is another way to evaluate scientific theories and it should come first, before picking up the first test tube. Simply, is the theory sound? Does it make sense in light of our established understanding of logic, the physical world and, in the current case, the properties of the biological cell?

For the vesicle theory, the answer is "no." This will no doubt come as a big surprise to many. Nonetheless, the fact remains that the theory fails on *prima facie* grounds. Even if all the mechanisms it proposes existed, they could not singly or together do what is required of them. It was this fact that prompted me some time ago to call the theory "incoherent".[74]

The deficiency lies in the requirement for balance, by which I mean not fairness, but correspondence or equality. The rate at which cells successfully make substances *must* perforce equal their average rate of transfer to the site or sites of their action, such as being secreted. For instance, and self-evidently, cells cannot secrete more than they make, and if they make more than they secrete, the presence of the substance in the cell would increase continuously and this, of course, is not compatible with a stable system. The same can be said for proteins that are not secreted, or for that matter, for any molecule manufactured by cells. Except for brief periods of time, they cannot be moved from the site of their synthesis to their site of action more or less rapidly than they are made.

This need for balance not only applies to the rate at which substances are moved *between* compartments, but their rate of deposition into and exit from them as well. Each step in the process must occur at a rate equal to that of synthesis, and as such, must equal each other. And this is not merely an operational requirement. It is fundamental and, as said, not only applies to proteins but also to all substances made by cells. Each step in the process of movement, whatever the mechanism or means, must equal the rate of synthesis. For instance, for the mechanisms imagined in the vesicle theory, every step from capturing the material in a forming vesicle, to the movement of the vesicle from place to place, to its fusion to a recipient compartment, to its release into that compartment, back to its budding from it, must equal the rate of synthesis.

This equality is absolute. It is not sufficient to say that the rate of transfer overall or that of any individual step is *proportional* to the rate of synthesis.[75,76] Though this is possible for vesicle and other similar mechanisms, mere proportionality will not do. The rates of the processes involved in movement must be identical, precisely equal the amount of the substance that is produced.

To the best of our knowledge, no physical process exists that can produce this strict equality when the conveyances are vesicles or any similar mechanism.[77] Equality between synthesis and movement can only be achieved between *the molecules themselves*, that is, by the direct chemical and physical comparison of their rates, as would occur if diffusion-based processes were responsible for movement. No intermediary, vesicles or otherwise, can serve as a substitute.[78]

# chapter twenty-seven

# The complexity excuse
## Avoiding Occam's razor

'Beauty is truth, truth beauty,'—that is all
Ye know on earth, and all ye need to know.

**John Keats**
*Ode to a Grecian Urn*

Keats' famous lines have been the subject of much analysis and controversy since they were written in 1819. What did he really mean by "Beauty is truth, truth beauty?" For that matter, what did he mean by "beauty," and what did he mean by "truth?" The urn of his ode was an elaborate vessel, complex both in its appearance and construction. So when he talked of the relationship between beauty and truth, he was not referring to some Platonic ideal of geometric simplicity, but to an intricate object.

He never explained the statement, so we don't know what he actually meant, but given the urn's ornate shape and intricate decorations, it is reasonable to think that at least in part it was in its complexity that Keats saw beauty and truth. This view of the beautiful is similar to graphic images of Mandelbrot fractal sets, simple mathematical truths that when iterated produce objects of great complexity, with a surfeit of curves, curlicues, and redundancies that are thought by many to be beautiful because of their embellishments.

Though I personally find Mandelbrot sets florid, not beautiful, and am not sure what I think of Keat's Grecian urn, we often see beauty in complex things. This brings me to the point of this peregrination—the most complex natural objects known to us are living things, and we often see beauty in them (as well as what is ugly and frightening). When contemplating a flower or the human form, it is not in the color of flowers or the geometry of the human body themselves in which we find beauty, but in their complex application, in their curves and curlicues. Just ask, is a red cube as beautiful as a red rose or the ideal human shape?

Nor does the complexity of living things end with such surface appearances. There is an elaborate world underneath, an underworld, found in the astounding complexity of protein molecules and, of course, the staggering brain. Indeed, the complex nature of living things, particularly that of higher animals and plants, is so breathtaking, so formidable that sometimes all we can do is wonder at it.

Given that we see beauty in extravagant complexity, it should come as no surprise that scientists trying to grasp the mechanisms of life have often gravitated towards complex explanatory theories and models. They seem necessary to reveal the truth. And it is in this light that there is often a *prima facie* conviction among biologists that it is in such complexity that we find nature's reality.

As a result, complex thoughts and proposals have a certain inherent authenticity. They seem lifelike. But there is an enormous problem with this way of thinking *for science*. Scientists are obliged, you might say duty bound, to set complex presumptions aside. As discussed, that ancient maxim of science, Sir William of Ockham's razor, issues a resolute warning to all who seek to understand nature to eschew complexity. Yes, living things may be very complex, but this is no excuse to ignore the dominion of Occam's razor.

In science, as opposed to art, complex explanations must be earned. It is not enough to say that what I see is complex and from this infer the truth of a particular complex explanatory model. Occam's razor requires the inquiring scientist to hold to the simplest explanation for natural phenomena *however complex things may appear*. Stated plainly, it says that the theory that explains a phenomenon most simply, that contains the fewest predicates, the fewest propositions is to be favored over those that contain more, and that as such are more complex.[79,80]

And so while complexity may be life's byword, science's is simplicity. The straits of passage between the two are perilous, and there is only one way to ensure a safe journey. Occam's razor must have paramount authority. It must rule. If we ignore its sanctions and let our imagination direct us to complex explanations, we may find ourselves forever lost, wandering tumultuous seas, never reaching safe harbor, all the while cleaving to the illusion that it is just over the horizon. So, we ignore Occam's razor at great risk. If we disregard it, what we believe to be true may be nothing more than a chimera, a mirage, and what we have exposed may not be nature's truth at all, but merely a reflection of our own passion.

Even worse, if we give complex ideas free rein, they may actually *obscure* the truth, while at the same time convincing us that it is in hand. We might think: how could such a complex system with so many explicit parts and elements be in error? There is too much here to be ignored. And that is the problem. If we allow ourselves the liberty of traveling down this path, if we depend on the presumptive truth of complex explanations, we need never abandon them however erroneous they may be. In this case, whatever challenges our theory may face, and regardless of reality, all we need do to continue to claim that it is true (or at least almost true) is add on whatever complexity seems to make it true.

Complexity is a soothing salve for the perplexed scientist. The most famous historical example of its use to force reality to fit a model, not the other way around, is the famous epicycles of Ptolemy. Discordant evidence was simply fit or squeezed into the view that the sun travels around the earth. This resulted in a complex theory that *almost* worked, but that turned out to be fundamentally in error, indeed, that had it backwards.

If we ignore the sanctions of Occam's razor, we necessarily cede ultimate authority in determining the truth to our imagination. This is unacceptable for science, where imagination, though critical, cannot be given free rein. If we allow it that authority, the result will be an unearned complexity, a complexity of the imagination, not one of nature's making.

This said, Occam's razor does not deny the existence of complexity. It merely requires that it be shown that simpler models and explanations fail to explain natural phenomena. It tells us that all available simpler explanations must be tested and rejected before contemplating more complex accounts.

In this process of testing, we are enjoined to move cautiously, one step at a time, from the simplest, to the less simple, and so forth, until a chain of rejections leads us inescapably to the least complex model able to faithfully capture nature's properties. Critically, in this progression, if we find that the simplest explanation is wrong, that does not give us license to jump hog-wild to any complex model of our liking even if we believe it to be true, and even if in the final analysis, it turns out to be correct. Its legitimacy must be earned one painful step at a time.

This being said, even physicists, chemists, and mathematicians who demand that Occam's razor be rigorously applied in their own field of study, become googly-eyed when contemplating the complexity of living things. They find complex explanations in biology, wondrous, beautiful, and pay little heed to how they are arrived at. This brings us to the vesicle theory and its conceits.

There are many examples of complex models in biology whose soundness has been earned by the evidence that supports them, such as the structure of the aforementioned protein molecules or the brain. But the founders and followers of the vesicle theory never sought to put their theory to Occam's razor's test. They liked what they had imagined, had some basis for thinking that it could be true, seemed to have good reason to doubt the alternative, and that was enough. With these beliefs firmly in hand, they set out to *develop* their theory, to elaborate its proposals, not test them, ignoring the precepts of Occam's razor in the process.

This is where my colleagues and I came in. The simplest explanation for the movement of molecules over short distances bar none is diffusion. As said, in diffusion-based processes, the molecules themselves provide the energy needed for their movement. The applicability of diffusion-based transport processes had never been tested for the digestive enzymes or for that matter for any protein molecule. As discussed, the reason for this was the preconceived notion that doing so would be pointless and the measurements worthless. The judgment that such processes do not occur was considered self-evident. What my colleagues and I did, our transgressive act, was to reject this presumption. We gave ourselves permission to carry out what many thought were senseless experiments.

And as I have outlined them here, what we found was astounding. The simplest (diffusive) modality accounted for what took place, while the complex vesicle theory, with all its contraptions and contrivances, failed test after test. The digestive enzymes seemed to move through and out of the cell and across its membranes on their own accord. In the end, whatever you think of the vesicle theory's propositions, even in the complex world of biology, Occam's razor must be respected. It must not only be given its proper due, it must be obeyed if the truth is what we seek!

# chapter twenty-eight

# Belief

## The beliefs of scientists

> But science can only be created by those who are
> thoroughly imbued with the aspiration toward
> truth and understanding. This source of feeling,
> however, springs from the sphere of religion. To this
> there also belongs the faith in the possibility that
> the regulations valid for the world of existence are
> rational, that is, comprehensible to reason. I cannot
> conceive of a genuine scientist without that pro-
> found faith.

> **Albert Einstein**
> *Science and Religion in Science, Philosophy and*
> *Religion, A Symposium, New York (1941)*

If you have escaped indoctrination or knew little, if anything, about the
vesicle theory before reading this book, and have not only been able to read
it with an open and curious mind, but also found yourself in agreement
with some, if not much, of what it has had to say, you must be marveling,
and are perhaps incredulous that the vesicle theory has been feted with
three Nobel Prizes. You may be puzzled that it is seen as settled science,
established for the ages, when it seems anything but. How did this come
about? How did the theory not only survive the weaknesses of the evidence on its behalf, as well as failed tests of its validity like those I have
outlined, to thrive, essentially unquestioned, untouched by any of it?

To find the culprit, we need look no further than the inordinate power
of belief. By "belief," I mean the acceptance of certain things as being true
despite being unproven, or if you will, unsubstantiated. At times, facts
upend beliefs, but unless or until that time, belief controls.

Many scientists, as well as the laity and political advocates of
particular science-based points of view, claim and probably truly think
that scientists do not harbor beliefs in unsubstantiated things. They think
that at least in their work they are exempt from this key feature of the
human mind. In science, reason and fact alone determine what is held to
be true. They would add that it is certainly not the other way around; with

unreasoning belief determining what scientists think is true. They would say that scientists only accept what has been substantiated by experimental or other valid test. They only trust in what has been proven. Mere belief holds no sway in their world.

And yet however appealing and however reassuring this thought may be, its sunny view of science is at odds with reality. The fact is that as they practice their profession, scientists are *not* exempt from belief in unsubstantiated things. Quite to the contrary, though often hidden or disguised, such beliefs are common, and are found in virtually every field of science. At times they concern important, even central concepts for which tests have yet to be devised, or, as the case for the vesicle theory, because the belief is so strong that testing the idea seems unnecessary.

The strangest thing about the beliefs that scientists hold is that they are *less* valid, less legitimate, and less defensible than religious beliefs. If you are modern and enlightened, perhaps an atheist, an agnostic, or at least a skeptic about various forms of religious belief, you might wonder how this can be. How can scientific beliefs be *less* valid than religious beliefs? Yet they are, and the reason is pretty obvious.

Ultimately, all religious belief rests on faith: either you have it or you do not. God's existence and the Holy Scripture being divinely inspired depend on it. Science, for its part, proudly proclaims that it has no truck with faith, that faith plays no part in its method and ministrations. Not even the clergyman Newton thought that his scientific discoveries were *based* on faith, even though he saw God's hand in them. The fact is that the absence of faith is central to what we call science and its exercise. Indeed, it is what originally set it apart from religion. Science is a faith-free zone.

But this poses a big problem. Belief, the adherence to unsubstantiated things, can *only* be a matter of faith, even if it concerns scientific subjects. As a result, whether admitted or not, in our time, as in the past, *faith* has always played an important role in science's understanding of the world. And yet, as said, a central tenet of science is that it does *not* depend on faith for understanding. It can even be said that its absence defines science.

So, it turns out that, unlike their faith affirming, faith-proclaiming religious brothers and sisters, when scientists and their supporters deny reliance on faith and yet hold various beliefs that are based on it, they are either confused or are hypocrites. As said, all beliefs, as opposed to facts, bar none, depend on faith, and there is no special exemption for science, or for scientific beliefs.

This does not mean that the scientist's faith resides in the overarching power of a deity. It may for some; though in this day and age, it probably does not for most. But whatever its basis, faith it must be, and as faith, it must come from somewhere and be based on something. If it is not faith in God, then in what does it inhere? It seems that the most prominent source today is solipsistic: faith in the power of the scientist's own mental abilities. Still, whatever the source, faith is required.

Nor is the insistence on denying faith-based belief in regard to scientific subjects a harmless conviction. If, to be dedicated to their profession, scientists cannot admit to faith as the basis for holding unsubstantiated scientific beliefs, we need to ask: "If not to faith, then to what can they be attributed?" Since there is no basis for belief other than faith, since it is faith or nothing, faith-denying scientists find themselves claiming that *their* particular beliefs, apart from all others, are not unsubstantiated at all, but are based on reason and fact even when this is patently false. They must either lie about the basis for their beliefs or deceive themselves as to their true nature.

The vesicle theory provides many excellent examples of this situation. Various beliefs about features of the model have been and are still claimed to be clear and unambiguous truths, when they are little more than the common beliefs of groups of scientists, not all that much different from the beliefs of a religious congregation. The problem is that, absent an attribution to faith, scientific beliefs become true by proclamation. Science becomes unyielding. Without faith in a particular belief, the believing scientist has no choice but to hold that what he or she believes is true by fiat.

Belief that the unsubstantiated is true provides a good working definition of a dogma. A dogma is something about which beliefs are held as facts. To this point, I have generally referred to scientific beliefs as "assumptions." Though my use of the word is correct, it gives undue credit to those who apply the term to features of the vesicle theory. It credits them with agreeing to or embracing the tentative and procedural character of their assumptions. It gives them credit for *not* believing that they are true, when quite the opposite is the case. The difference between a belief and an assumption is that the latter is accepted as being true without proof as a matter of convenience, as in mathematical proofs, while the former is held to be true without proof as a matter of faith, however strongly denied.

Even though one's personal beliefs may not affect events in the outside world, they serve as the basis for all our actions; indeed, they serve as the basis for all actions among thinking animals other than automatic neuronal reflexes. These beliefs may be positive. For instance, belief in a certain affirming outcome may enable perseverance. Or they may be negative, even to the point of death and mayhem. A dramatic and evil historical example among humans has been the millennial-long persecution of the Jews.

As it relates to science, on the positive side of the ledger, holding belief in an idea can give us the courage to persist in our efforts to understand nature. On the negative side, it can not only lead to unwarranted confidence that a particular belief is true when it is not, but that those who do not hold it are wrong on this basis alone, on the basis of our confidence in the belief.

In science, if such a belief is widely held, that is, if it represents the standard point of view, it can provoke the conviction that we have an obligation to rid ourselves of the wrongheadedness of nonbelievers. After all, the task of science is to stamp out ignorance, to liberate humanity from inaccurate, fallacious, and baseless thinking. As such, those who hold dissident or heretical views may not only be seen as ignorant, but treacherous. They must be prevented from promulgating their false and misguided ways of thinking as well as their erroneous or fabricated observations.

In trying to silence nonbelievers and purge their pernicious ideas, believers may be acting in good faith. They might say that all they are doing is attempting to cleanse science of defective or dishonest thinking. And who can fault them for that? Even if they are misguided, their motives are honorable. But not all who seek to purify science in this way are so nobly motivated. The actions of the less than principled fall into three categories—expressions of anger, acts of cowardice, and the desire for power, or some combination of the three.

Anger is often expressed in unbridled contempt for the transgressor. *How dare you? Your views endanger our hard-earned understanding. You and your ideas must be squelched, censored, in a word silenced.* The cowards are those of weak will or mind who seek to enforce the standard belief for fear that they will be seen as *not* supporting it and be branded as being in opposition. As the saying goes, they go along to get along. For some, it is not fear of being tarred as a transgressor that is at work, but it is simply their nature to follow the leader. They are members of "the herd of independent thinkers." Finally, there is the third reason—those who seek power and control, who hide their base motives in the sweet language of science.

In all this, belief is the strong horse, and reason and evidence the weak horse. As a young man beginning my career in science, I held two thoughts uncritically. The first was that in science, unlike life otherwise, nature's truth always determines belief. Second, its purveyors are committed to that truth, both as individuals and in their institutions. I discovered far too slowly and with much anguish that neither is true. The devotion I confronted was to a model, not to nature.

For some, allegiance to a belief wholly defines their careers in science. They express its ideas and their dedication to them in everything they do. Their fealty is proclaimed proudly, with attendant hosannas in various journals and in the activities and efforts of learned societies. This can be seen in what journals publish, what scholarly symposia various societies sponsor and whom they honor. Agreeable data are greeted with cheers and little criticism, and pass peer review quickly and unfettered. Whereas data that are disagreeable, that question or weaken the paradigm are viewed harshly, publication of the research and ideas commonly being delayed or rejected.

The same devotion can be seen in whom universities hire, in who is afforded endowed professorships and large laboratories, and conversely in who is ignored or rejected, relegated to low rank, back rooms, and cellars. Likewise, in regard to government and foundation funding, there is often a cloying aggrandizement about how "creative" particular research is, when what is being fostered, what is being favored may not only be predicable and pedestrian, but unquestioning, abiding by, and promoting the common view.

None of this should be surprising, though it certainly surprised me as a young man. It is simply human. My experience was not a one-off happening. Nor is my criticism reflective of no more than a personal gripe. This is science at work, no less today than in the past. As Kuhn claimed, paradigms and their subsumed beliefs dominate most scientific activity and challenging them is never appreciated.

Finally, it is important to keep in mind that beliefs are, by their very nature, unsubstantiated. To say that something is an "unsubstantiated belief" is redundant. To say that it is a "substantiated belief" is an *oxymoron* and *non sequitur*. Substantiated beliefs are facts, and hence are not beliefs at all. On the other hand, to call beliefs facts, when the appellation does not apply, is to hold to a dogmatic creed that does a disservice, sometimes a great disservice to science.

# chapter twenty-nine

# The rise of the careerist
## Science as a profession

> No science is immune to the infection of politics
> and the corruption of power. ... The time has come
> to consider how we might bring about a separa-
> tion, as complete as possible, between Science and
> Government in all countries. I call this the dises-
> tablishment of science, in the same sense in which
> the churches have been disestablished and have
> become independent of the state.
>
> **Jacob Bronowski**
> *In The Disestablishment of Science,*
> *Encounter, July 1971, 15*

As a young man barely out of adolescence, looking for a career of some
sort in one field or another, what I found was not a career, but a calling.
After serious missteps, I had settled on my heart's desire. I was going ask
questions of nature. Come hell or high water, in one way or another, I was
going to be a scientist.

Throughout most of its history, science has been more of a calling
than a career. The wealthy could dedicate themselves to their passions
without the need for financial support, while those without the resources
had careers in other fields to pay the bills so they could pursue it. They
were physicians, members of the clergy, or worked for the bureaucracy or
the crown. From Galen the physician to Einstein the patent officer, their
callings, their passions, and their jobs were different things. Even university professors made their living by teaching, not doing scientific research.
Whatever your career, science was an avocation, an adored hobby for the
few so disposed or so possessed.

In this, I am talking about scientific inquiry for its own sake, not about
the application of scientific principles to meet practical needs, so-called
"applied science." From its very beginnings, science's practitioners sought
to apply scientific principles to utilitarian needs, to develop more effective weapons of war or instruments of civic betterment, such as aqueducts
and canals. These efforts were not only promoted by the state, they were

promulgated, decreed, and even implemented by it. During the industrial revolution the need to apply science to achieve practical aims grew enormously, and as a consequence, so did government involvement. In the United States in the mid-19th century, the U.S. Congress formalized its participation with passage of the Morrill Acts, which led to the creation of state colleges and universities, so-called "land grant colleges." Their objective was to provide a scientific education for students in utilitarian subjects, like agriculture and engineering, as well as carry out research applying scientific principles to these subjects.

Almost a century later, by the mid-20th century, when I was training for a life in science, almost all institutions that carried out scientific research were wrapped in the warm and supportive financial embrace of government. Almost all science was supported by public money and taxes. With this backing science became a *career*, a full-throated substantial, valued, and influential job, and the notion of a "calling" became an anachronism reserved for clerics.

One no longer needed wealth or another job to sustain your interest in science. With sweeping government subsidies for training and research, there were more active scientists in the world than all those who had come before. Government largesse had greatly expanded what was now the *profession of science*. Along with this support came a career at a university or other research institution with good pay, promotions—not to mention, prestige and power.

No longer was a life in science merely an expression of one's personal curiosity about nature. It was now an esteemed career. Its practitioners were seen as the brightest among us, and their opinions were granted the highest respect. At the same time, "pure" science, science for its own sake, though still fodder for TV science programs, increasingly came to be seen as irrelevant, unneeded, though charming and interesting, a disposable indulgence and a waste of precious time and money. Whatever one thought of the pursuit of the esoteric, science was no longer an arcane discipline reserved for talented dilettantes and enthusiasts. It had become a technical career intended to meet the practical needs of society.

It was widely understood that the scientific approach could be of great benefit to society and that, as such, it should be applied to meet valuable practical goals. This made science an important governmental responsibility. In the unprecedented mobilization of scientists during World War II to develop instruments of war, such as breaking codes and building bombs, government provided the means and direction, and scientists carried out the task. They were servants of society's needs. As a result, and with the growing application of science to medical and public health needs, the distinction between science as a means of learning about nature, undertaken without practical purpose or intent, and its use to achieve utilitarian goals was blurred.

It was no surprise then that when the war ended, things did not return to the *status quo ante* with government and science returning to their respective corners with different interests and concerns. To the contrary, their partnership grew enormously in both scope and size. To borrow Eisenhower's phrase, in the years following World War II, a *university/government complex in science and technology* was forged with the object of fulfilling major societal goals. In creating this alliance, it was appreciated that curiosity could be useful, and this justified government support for fundamental research about the nature of our world. It was believed that even the most basic scientific research would sooner or later be of some value to society. Basic science was no longer just an expression of the enthusiasms of the curious, no longer was it just a matter of personal inquisitiveness; it was there to serve.

This utilitarian goal of the government/science compact was a reflection of a broader, utopian political and social agenda. Whatever form it took, and it has taken many, government and its bureaucracy were seen as the solution to society's problems. Give us the resources, the taxes, the ruling class said, and we will hire and train people in various fields who, through their expertise, will better our lives.

Science was an essential part of this paradisal plan. Scientific experts would be trained and then drafted to serve the needs of society. The physical sciences would ease life's burdens by developing new machinery and new technology, while research in the biological sciences would cure disease and heal the planet. However abstruse, science had become a technical career whose purpose was to meet the practical needs of society.

Given this goal, it is not surprising that the state's interest was not limited to providing funds and cheerleading the effort. Science was far too important a business to be left to the whims of wooly-headed scientists. It had to be guided and controlled to prevent its descent into a kind of scientific solipsism, a navel gazing in which the purely intellectual interests of scientists, divorced from the practical needs of society, would dominate.

A central consequence of the compact was that science and scientists, not to mention their employers (mainly the university), not only became wards of the state but also the state set the agenda. Based on government support, universities became behemoths of money, power, and prestige. If that support were to disappear tomorrow, so too would the modern research university. Its laboratories would close and its giant administrative structure would collapse (no doubt last).

Though societies had underwritten giant engineering projects for thousands of years, the infusion of large sums of government money made it possible for the first time in human history to carry out huge *scientific* projects far beyond the capability of small laboratories and individual scientists. It was hoped they would yield advances comparable to their size and scope. Giant accelerators were built to break atoms into smaller and smaller pieces to learn more about the intimate structure of matter, and the Genome Project enumerated the genes of humans and many other species, transcribing nothing less than "The Book of Life." History will decide whether the bang has been worth the buck, but the scientific work that took place as part of these gigantic projects, as well as apart from them, in small laboratories, were the product of aims in great part set in Washington, DC.

# chapter thirty

# Cronyism

## Politics in the world of science

Academic politics is the most vicious and bitter form of politics, because the stakes are so low.

*Usually attributed to Henry Kissinger, though the correct attribution is probably Wallace S. Sayre (1905–1972), Professor of Political Science at Columbia University, The Wall Street Journal, Dec. 20, 1973. (For the thought's origin, see the Samuel Johnson epigraph to Chapter 23)*

This is not to say that Washington bureaucrats based their decisions solely or even mainly on their own judgment. They understood the need for input from scientists, from experts in particular subjects drawn from great universities and research institutions. To obtain their advice, the government invited scientists to serve on special advisory committees and other bodies of notables that were to determine what subjects were worthy of study and what research was likely to be fruitful, and in this light, which areas should have a high priority for funding. The government then mirrored these opinions in what it mandated. With advisory committees having laid out what to study, other committees of experts then provided peer review for particular research proposals in those areas.[81]

This made perfectly good sense. Who else should determine the direction of scientific research if not scientists in the particular field, if not the experts? Their having a major role in setting the direction of government-supported research, as well as in evaluating particular proposals not only made sense, it was essential and not merely in a technical sense. Their involvement would remove the potential taint of politics from the government's decisions. The assessments of advisory and peer review committees would be scientific—objective, not political.

And yet, predictably, the involvement of expert scientists in the review process did not guarantee objectivity, far from it. As said, experts come with built-in agendas, attitudes, proclivities, and desires, both as individuals and in groups. But far more disturbing, those who served on the peer review and advisory committees, with the government behind them as enforcers, became de facto arbiters of correct thinking in the particular area.

Beyond this, the choices of the advisory committees were often so carefully tailored that the possibility of the research being carried out by anyone other than members of a select group, notably committee members themselves [Note: recent funding scandals in the Environmental Protection Agency (EPA)] unlikely. Hidden or unabashed, cronyism was rampant. A select group of scientists acted to secure their own interests under the pretense that they were the same as those of science. Critically, in ordaining what scientific work was worthy, competing or conflicting ideas were, by default or exclusion, deemed unworthy.

Nonetheless, despite such terrible weaknesses, this system, particularly government-sponsored peer review of research proposals, became so influential that it often usurped the traditional review processes used by universities to determine whether a particular faculty member's scholarship was or was not meritorious. Committees set up by government, not department chairs, deans, or faculty became the arbiters of scientific accomplishment. Many saw this as a good thing, not an unwelcome intrusion. It liberated the evaluation of faculty from base political forces at work within the university. It replaced the parochial infighting that often left an individual's promotion vulnerable to personal animosities with an objective evaluation.

If a professor's laboratory had government grants, it meant that the scientific community at large, throughout the country, indeed throughout the world, thought well of them and their research. In this system, government-sponsored peer review and the subsequent funding of research grants was the marker of intellectual accomplishment. In it, only the science mattered. The fierce internecine politics of university life was supplanted by an unbiased and objective evaluation of the ideas and work. If outside peer review panels deemed the research worthy, the faculty member was praiseworthy and deserving of promotion. If they thought the research unworthy, if funds were not granted, this attested to the fact that the professor and his or her research was substandard and that promotion was not warranted. And most wonderful of all, this determination was impartial. Judgments were rendered without privilege or favor.

The central consequence of this new order was that without government support, the experimental scientist was most often unable to carry out significant scientific research. There was a kind of *ostracism in practice*, if not theory, exemplified by the failure to garner government support. Without it, as said, there was no research, and without research, there are no publications, and as the saying goes, without publishing, academics perish. Still, however unpleasant for those who were rejected, what could be fairer, more considered, more dispassionate than peer review by outside government panels? Scientists, experts in specific research areas, not just your neighbors, evaluated the character and quality of your work.

Even so, unsurprisingly, the new arrangement was far from perfect. There was bias, favoritism, and prejudice aplenty, just expressed in a different context. Whether at the broad policy level or in the details of particular research proposals, the process was rife with special interests, special pleading, and foreseeable corruption. Yet despite its flaws, it seemed like a major improvement over the old order. Yes, it introduced prejudices and biases, but they were less personal. They were more scientific.

As difficult as this arrangement can be for those who labor inside the lush vineyard of "normal" science (and it can be difficult) the system works well enough. The fact is that for those engaged in Kuhn's normal science where support for the paradigm is sought, and where there is general agreement among scientists about what they are doing and where they are going, and putting personal animosities and the formidable ego-driven character of many scientists aside, and however diverse and conflicted their approaches, they are trying to reach the same, shared goal. They are on the same quest.

But this is cold comfort to those laboring outside the reassuring caress of the paradigm, to those who harbor a fundamentally different opinion, who for one reason or another find themselves questioning the common belief. For them, the system fails and fails miserably. In their hardscrabble world, government-instituted peer review is not merely unfriendly; it is the enemy. Its processes are usually greatly and often terminally skewed against their dissonant point of view.

How can an aberrant or nonconforming perspective get a fair hearing in a system designed to promote conformity, not skepticism? Indeed, committees of experts are likely to be nonplussed, if not outraged by deviant ideas and the data supporting them. "What you are saying runs counter to everything we know to be true." "What you suggest is inconsistent with well-established understandings and is unworthy of support, either intellectual or financial." Though of course speech is still free, and you can always stand on a soapbox and try to attract interest, the offending line of thought and research, however objectively praiseworthy, will in all likelihood not take place.

Not to diminish the problem faced by such individuals, the question for serious contemplation is not their bruised egos or even the damaged careers (not that that is irrelevant), but the effect that this has on scientific progress. How likely is it that true (opposed to faux) scientific breakthroughs will emerge in these circumstances? In liberal societies such as ours, where those who propose uncomfortable ideas are not incarcerated or executed, punishment is still meted out by government-chosen committees of experts. Like the Popes or Potentates of old, they work to extinguish heresies.

Modern science is at base a collectivist enterprise in which the truth is determined by the votes of committees of scientists who harbor common beliefs. This may well be a great improvement over the personal opinions of kings and clerics, but it is no intellectual paradise. Committees of experts tend to diminish, to discredit, and most importantly, to exclude opinions that question their shared view. Such opinions are considered misconceived. The result is that to achieve professional success in modern science almost always requires hewing faithfully to one or another "company line." It requires conformity to and compliance with the paradigm.

When historians tell us of the checkered past of scientific work, with its epicycles, alchemy, and myriad fanciful imaginings about living things, we smile with a contented condescension. We are so fortunate. We understand so much more today, and what we know, we know securely. Yes, many details have yet to be fleshed out, and yes sadly, the giant questions of origins may remain forever beyond our reach, but look at what we have achieved. Our understanding of nature is substantial, safe, and sound.

But if we look with a skeptical eye, this complacency seems guaranteed not so much by reason, as by the democratic ideal of peer review. It is peer review that gives us confidence that modern science is untainted by the idiocies of the past. But this means that what we think we know to a certainty is born of the power of group opinion, determined by a vote of peers, not logic and reason.

Whether or not we are willing to admit it, the direction and supervision of most modern research resides in the hands of a small chosen scientific elite who are given the task of defending and strengthening the paradigm. In this enterprise, independence of thought and action, though invariably applauded in theory, is, in fact, discouraged. In the era of the careerist, many scientists function as high-class technicians, cogs in someone else's intellectual endeavor. They are the very smart "worker bees" of modern science.

One curious outcome of this system is that small grant proposals—say, under a million dollars a year—the provenance of a single investigator often receives far greater scrutiny, at times down to the details of individual experiments, than large projects whose costs run to hundreds of millions, even billions of dollars that are supported by communities of scientists who fly the same banner and advocate strenuously in its behalf.

There is no way around it: in this system, the independent mind is greatly disadvantaged, and the common view, even if dead wrong, is reinforced. While no doubt there are individual scientists out there who quietly harbor revolutionary thoughts and are carrying out innovative, world-shattering research in the hidden nooks and crannies of our universities and research institutes, they breathe clotted air.

No doubt many would agree with me about the flaws in the current system, but would add that despite its flaws, in the final analysis, it is enlightened and progressive, and eventually open to new ideas. Yet, if one looks with eyes wide open, the system is closed and reactionary by its very nature. Good things may happen, but in spite of not because of the system. Whatever the subject, whatever the field, ideas and evidence that challenge the standard view are without fail examined skeptically, with the presupposition that they are wrong. While ideas and evidence that comply with the standard view are not only greeted leniently, but also often with enthusiasm and satisfaction, regardless of their veracity. And though discoveries that truly change our understanding of nature have always been, and no doubt will remain rare, those that successfully challenge the solidified world of the paradigm are rarer still.

From time to time, I shared this negative opinion with colleagues. Invariably, they challenged me with a single question: "Do you have a better idea, a better way of making these determinations?" Though I thought that there were demonstrably better ways, the point of the question was not to elicit ideas for reform. Just as there was only one question, there was only one correct answer. Whatever its flaws, whatever the abuses, this was the best we could do and we just had to live with it. There is no better system, no less one that is magically and truly fair and unbiased.

# chapter thirty-one

---

# Psyche
## Why me?

> "... there is no substitute for a careful, or even meticulous, examination of all original papers purporting to establish new facts."
>
> **R.A. Fisher**
> *Has Mendel's Work Been Rediscovered?*
> *Annals of Science, 1936, pg. 137*

"Why me?" is not a complaint about mistreatment. Rather, it asks what character or personality trait led to my becoming a skeptic and iconoclast. All humans leave the warm, soft world of the womb to enter the great cacophony that is life after birth. From its safety, we are pushed unprotected into the dangerous outside world.

In the early days, all we can do is cry for our mother's help. As we grow out of infancy, we develop all sorts of stratagems to protect ourselves in the intimate world of the family. But before long, we find ourselves having to adapt to the world beyond. In our efforts to secure our wellbeing, we make use of traits common to us all, as well as those that are the unique result of our personal experience. Add the unpredictable events we call fate to the brew, and our personality, our character, who we are and how we get along in the world, emerges. Driven by the unquenchable desire to survive, we become optimistic or pessimistic, hopeful or lacking hope, confident or insecure, sure or irresolute, accepting or rejecting, thoughtful or reactive, direct and forthright, bent and twisted.

As any parent knows, children display two key traits that exist in painful opposition to each other. Our children are, at one and the same time, naïve and willful. Naivety is lost slowly as they experience life's many difficulties. During adolescence, this becomes a full-on crisis of faith towards parents, family, and eventually the larger world. With the loss of naivety, willfulness stands alone. But it, too, is doomed. Even as young children, and much against our will, we succumb to our parents' wishes and comply. Subsequently, most of us in one way or another submit to demands imposed on us by the outside world. We obey our teachers and eventually the laws of society.

With the loss of naïvety and willfulness, we come to express their opposites, cynicism and compliance. Though many, if not all of us, try to one degree or another to resist these shifts in attitude, some are more successful, if that is the right word, than others. Some are able to retain a childlike naivety, while others hold on to their independence through willfulness. A few continue to display both naivety and willfulness borne of childhood as central features of their adult life.

I believe this has been my destiny. To one degree or another, and for good and ill, I have held on to both traits of character. They became vital features of my life, and in regard to the subject of this book, of my life in science. Only ever so slowly, with large doses of heartrending experience, did I become aware of my naivety in regard to the motives of others and my obstinacy in insisting on charting my own path despite the attempts of those in authority to force me into line.

I tried to be compliant, and often thought that I was, but just as often I failed. My naivety and willfulness were deeply rooted. As I have outlined them here, they have, at times, made my life in science difficult, even painful. After a long life, I sometimes think that I have mastered or at least tamed these twin impulses. I try to convince myself that at last I have become more realistic, only to find that nothing has changed. To both my benefit and detriment, both fortunately and unfortunately, I have remained naïve about the attitudes of others and willful in my assertion of independent thought. This way of being has not been not wholly without its blessings. In my life outside science, I have found the meaning of love and devotion. In science, it has given me the great joy and exhilaration of discovery. It has allowed me to question nature about her wonders directly, without obligation or preference. Compliance, however well rewarded, can never match that.

## chapter thirty-two

# Carved in stone
## Dogma

> Whoever undertakes to set himself up as a judge of
> Truth and Knowledge is shipwrecked by the laugh-
> ter of the gods.

**Albert Einstein**
*Essay to Leo Baeck (1953)*

We have talked about one difference between the creative impulse
in art and science, but there is another that is no less significant. Once
Michelangelo stopped working on a statue, whether he was satisfied
with his accomplishment or disturbed by his failure, it was finished,
done. Save for the ravages of time and human mishandling, the object
he produced was permanent, unchanging, carved, as it were, in stone.
And when Botticelli decided that his masterpiece *Primavera* was finished,
its forms and figures, their juxtaposition to each other, all the colors and
shadings of the painting, were set on the canvas for the ages. So it was
for a Mozart sonata or a Chopin *Prelude*, on and on, artistic endeavor by
artistic endeavor.

By contrast, *nothing* in science is carved in stone; nothing is permanent. Everything is vulnerable, provisional. Scientific wisdom, however
well established, is always up for reevaluation, for reconsideration. The
absence of faith aside, the impermanence of its understandings can be
said to define science. Scientific insights, however seemingly indubitable,
can always be overturned by new knowledge, by new understandings.

Or at least this is how science is thought to be. However, in practice,
like all other systems of belief, its understandings *can* be set, statue-like,
in stone, not open for reconsideration or even discussion. These are the
dogmas of science and they give it an air of permanence, just like a statue.

In 1958, Francis Crick coined the term "the central dogma" for the
understanding that DNA gives rise to proteins by means of an irreversible chemical process. This was meant as a compliment, a declaration that
the idea was true forever. But whatever his intentions, to call any scientific
understanding a dogma, to say that it is true forever, is not a compliment.
It is to place it beyond science and just as surely, on that grounds, beyond

reason. Though not intended, to call it a dogma was pejorative. It was to call it nonscientific even though the idea was grounded in science.

When scientific principles are thought to be incontrovertible and forevermore true, they are no longer scientific principles. Someday, perhaps in ways we cannot yet imagine, the incontrovertible may be shown to be false. Not only are such occurrences conceivable, the history of science is littered with them. In fact, it can be argued that the rejection of entrenched opinion is the basis of all fundamental advances in scientific thinking.

As should be obvious by now, unlike Crick's statement about the central dogma, my calling the vesicle theory a dogma is not meant as a compliment. To the contrary, it is intended to question its claims of authenticity and authority. As we have here, when we look back, we find that much of the affirming evidence was weak; almost none of it quantitative, and some purported support for the theory was no such thing. And this is not to take into account, failed tests like those I have laid out. Still, despite all this, the theory sits at the very center of the contemporary view of the biological cell, allegiance to it indeed set, like *David*, in stone.

As we have discussed, unlike scientific dogmas of the past imposed by the church or the crown, the modern dogma is arrived at democratically. It is an expression of the shared beliefs of communities of scientists, of experts in a field. With time, their consensus spreads to the broader scientific community, and eventually to the laity. It dictates which research will be carried out, which will be published, and which will not, who will be promoted and who will be honored, and most frighteningly, what is understood to be true and what is understood to be false.

The truth, it seems, is up for a vote, a matter of groupthink. Whatever the group thinks is de facto true regardless of the facts. If a community of experts and certainly if the scientific community at large believes that X is true, then X is true. Any contrary or doubting opinion is false. The contrarian is thought a fool or a scoundrel, but either way is wrong.

This is not to say that scientists do not understand that reason and proof, not majority opinion, determines what is true and what is false. Most do. But they do not see the power of majority opinion as a problem. To begin, it is, they fairly surmise, probably correct. Numbers *do* count. But if it is wrong, not to worry—it will eventually be overturned, dislodged, and corrected in the ordinary pursuit of scientific knowledge.

I believe that this conviction is little more than comforting balm. Thomas Kuhn argued that though logic and evidence can play a role in overturning scientific systems, the toppling of paradigms and their attendant dogmas only occurs if and when scientists lose confidence in it, and this, Kuhn avers, has little to do with logic and evidence, and much to do with the social and political world in which the change takes place.

In telling the story of my failed attempt to challenge the powerful vesicle theory, I have exposed some of its weaknesses and described various experimental tests that it failed. It was in this light that I argued that however deeply embedded in the scientific Zeitgeist, the vesicle theory is false. It simply does not comport with the properties of nature. Not only that but its failure, exposed a vital and previously unknown biological process. Protein molecules, it turned out, can cross biological membranes individually according to the laws of diffusion.

But I have also explained none of this mattered. However damning, the controlling authority of the cherished belief was never weakened by our observations or anything we said. Not only was there little to no introspection, but the opposite occurred. There was a reaffirmation, an insistent and unwavering assertion that the truth, permanent and unvarnished, was in hand and known. It is fair to say that the paradigm and its vesicle theory are more strongly embedded today than ever. And though the occurrence, indeed the importance of the movement of fully formed proteins across membranes individually is widely appreciated today, due to the vesicle theory's dominion, and however odd it may seem, this is still forbidden for all proteins manufactured on attached ribosomes.

There is much hand-wringing nowadays among educators about the importance of teaching students "critical thinking." They complain that their students do not know how to analyze the world of ideas in a thoughtful and independent way. They simply parrot the teacher's, that is to say, the authority's point of view.

This problem is invariably discussed in reference to the humanities, but it applies to science as well. Students of science are taught, *qua* indoctrinated in established thought in both textbooks and lectures. Scientific principles are commonly presented as irrefutable, as manifest facts, not to be toyed with, but wondered at and viewed with awe. Only rarely is the idea floated that however widely held and honored, some of these beliefs are probably wrong, even if we do not know why or in what way. Richard Feynman's writings and public lectures about physics stand as a noble attempt to expose students and the public to the uncertainty of scientific knowledge. I am unaware of any comparable exposition in biology. Most often ideas in biology, ideas about molecules, chemical reactions, cells, organs, organisms, and even communities of organisms are presented as facts established for the ages, as certain as the rising sun.

I remember a senior colleague at UCSF, a self-avowed iconoclast, giving lectures to students about how nerves work. I sat spellbound as he described how our modern ideas came about. He exposed the students to the actual experiments and data, much of it dating back to the 19th century. He not only told them where our ideas came from, but he also exposed them to the experimental and theoretical bases of those ideas, both the understandings and the uncertainties. Though his lectures enthralled me unfailingly year after year, he received by far the poorest ratings from the students. When asked what the problem was, they said in seeming unison that they did not need to know about all these experiments. They just needed the facts. They were very busy, indeed overwhelmed by their workload, and they simply did have not the time to figure it all out.

I have outlined experiments we performed that tested the vesicle theory in much the same way that my colleague talked about nerves in his lectures. I have tried to describe the results of actual experiments and the meaning we imputed to them, as well as juxtapose the interpretation of scientists who believe in the vesicle theory.

Still, the central lesson of this narrative is not found in these tests or their meaning. Indeed, it is not found in the truth or falsity of any theory or any particular understanding. It is coming to appreciate that the greatness of science is *not* to be found in what we know or think we know about the world around us, but in our awareness of the tentativeness of our knowledge.

It is scientists who carry out their mission as critical witnesses to their own certainties, as well as the various shibboleths of their field, that uncover nature's truths. Those who try to impose their views, biases, and attitudes not only on other scientists, but on Mother Nature herself, can only give us a twisted and garbled view of the world. They provide a landscape where false knowledge masquerades as true understanding. They replace the meekness of the supplicant towards nature by human arrogance and its boon companion ignorance. And without humility, with certainty replacing modesty, we end up with dogma.

Just as true religious believers prostrate themselves before God, to be true to their craft, scientists must humble themselves before the mysteries of nature. To do this, they must admit their ignorance and their unsteady grasp of the world they study. Only from such a diffident point of view, only by putting beliefs and biases aside, can we listen attentively and adroitly as nature speaks. The magnificence of the human's unique ability to learn about the properties of the natural world can only be fully realized when we ask nature to reveal herself, and in taking note, take special care not to overwhelm her with our beliefs. As her minions and stewards, we must open our minds and hearts to the answers she provides, and not seek to impose our own.[82-98]

# Notes and selected readings

## Chapter 2

To learn a little about Professor Heilbrunn and his research read:

(1) Chapters 7 and 8, *A Real-Life Parable* and *The Lesson* in Stephen Rothman, *Lessons from the Living Cell* (McGraw-Hill, New York, 2002).

## Chapter 3

(2) Rothman, S.S. Exocrine secretion from the isolated rabbit pancreas. *Nature*, 204: 84–85, 1964.

(3) Rothman, S.S. and F.P. Brooks. Electrolyte secretion from rabbit pancreas in vitro. *Am. J. Physiol.*, 208: 1171–1176, 1965.

(4) Rothman, S.S. and F.P. Brooks. Pancreatic secretion in vitro in "Cl–free," "$CO_2$–free" and low Na+ environments. *Am. J. Physiol.*, 209: 790–796, 1965.

## Chapter 4

(5) Babkin, B.P. *Pavlov: A Biography* (University of Chicago Press, Chicago, 1949).

For Babkin on parallel secretion see:

(6) Babkin, B.P. *Secretory Mechanism of the Digestive Glands* (Hoeber, New York, 1950).

## Chapter 5

My favorite book of electron micrographs on the fine structure of cells from this period is

(6a) Fawcett, D.W. *The Cell* (W. B. Saunders, Philadelphia, 1981).

The images are beautiful, but as with so much of this literature, they have been carefully prepared and selected—you might say manicured—and as such, are as much a reflection of the artist's eye and choices, as the appearance of the forms as they are found in nature.

## Chapter 7

Heidenhain's original paper on the disappearance of zymogen granules is

(7) Heidenhain, R. Beitrage zur Kenntis der Pancreas, *Pfluger's Arch.*, 10: 557–632, 1875.

For granule shrinkage, see

(8) Kühne, W. and A. Sh. Lea. *"Beobachtungen über die Absonderung des Pankreas"*, Untersuch. physiol. Inst. Univ. Heidelberg, 2, 448, 1882.

## Chapter 9

The first review outlining research on the vesicle theory by some of its creators is

(9) Palade, G., P. Siekevitz, and L. Caro. Structure, Chemistry and Function of the Pancreatic Exocrine Cell, *Ciba Foundation Symposium on the Exocrine Pancreas: Normal and Abnormal Functions*, eds. A. deReuck and M. Cameron (Churchill, London, 1962).

## Chapter 10

The original research papers providing evidence to support the vesicle theory are

(10) Siekevitz, Palade and Caro. *J. Biophys. Biochem. Cytol.*, In volumes 2–7 from 1956–1960.

And several years later,

(11) Jamieson, J. and G. Palade in the *J. Cell Biol.*, In volumes 34, 39, 50 from 1967–1971.

For the study with gold beads, see

(12) Kraehenbuhl, J., et al. Immunocytochemical localization of secretory protein in bovine pancreatic exocrine cells. *J. Cell Biol.*, 76: 406–423, 1977.

## Chapter 11

(13) Redman, C.M. and D.D. Sabatini. Vectorial discharge of peptides released by puromycin from attached ribosomes. *Proc. Natl. Acad. Sci. USA*, 56: 608–615, 1966.

(14) Redman, C.M., P. Siekevitz, and G. Palade. Synthesis and transfer of amylase in pigeon pancreatic microsomes. *J. Biol. Chem.*, 241: 1150–1158, 1966.

(15) Blobel, G. and B. Dobberstein. Transfer of proteins across membranes. I. Presence of proteolytic processed and unprocessed nascent immunoglobulin light chains on membrane-bound ribosomes of murine myeloma. *J. Cell Biol.*, 67: 835–851, 1975a.

(16) Blobel, G. and B. Dobberstein. Transfer of proteins across membranes. II. Reconstitution of functional rough microsomes from heterologous components. *J. Cell Biol.*, 67: 852–862, 1975b.

(17) J.J.L. Ho in S.S. Rothman. *Protein Secretion: A Critical Analysis of the Vesicle Model* (John Wiley and Sons, New York, 1985).

## Chapter 12

For two contemporaneous critiques of the signal hypothesis, see J.J.L. Ho and S. Rothman in

(18) *Nonvesicular Transport* (John Wiley and Sons, New York, 1985).

## Chapter 13

(19) Popper, K. *The Logic of Scientific Discovery* (Basic Books, New York, 1959).
(20) Kuhn, T. *The Structure of Scientific Revolutions* (University of Chicago Press, Chicago, IL, 1962).

## Chapter 14

(21) Rothman, S.S. "Non-parallel transport" of enzyme protein by the pancreas. *Nature*, 213: 460–462, 1967.

The same phenomenon a decade later:

(22) Rothman, S.S. and H. Wilking. Differential rates of digestive enzyme transport in the presence of cholecystokinin-pancreozymin. *J. Biol. Chem.*, 253: 3543–3549, 1978.

Also, for the distribution of digestive enzymes in zymogen granules see:

(23) Bendayan, M., J. Roth, A. Perrelet, and L. Orci. Quantitative immunocytochemical localization of pancreatic secretory proteins in subcellular compartments of the rat acinar cell. *J. Histochem. Cytochem.*, 28: 149–160, 1980.

For an early review on non-parallel transport see:

(24) Rothman, S.S. The digestive enzymes of the pancreas, a mixture of inconstant proportions. *Ann. Rev. Physiol.*, 39: 373–389, 1977.

Nonparallel secretion 25 years later:

(25) Rothman, S.S., C. Liebow, and J. Grendell. Non-parallel transport and mechanisms of secretion. *Biochim. Biophys. Acta*, 1071: 159–173, 1991.

## Chapter 15

(26) Liebow, C. and S.S. Rothman. Membrane transport of protein. *Nature (New Biol.)*, 241: 176–178, 1972.
(27) Liebow, C. and S.S. Rothman. Equilibration of pancreatic digestive enzymes across zymogen granule membranes. *Biochim. Biophys. Acta*, 455: 241–253, 1976.
(28) Liebow, C. and S.S. Rothman. Transport of bovine chymotrypsinogen into rabbit pancreatic cells. *Am. J. Physiol.*, 226: 1077–1081, 1974.

(29) For a review on protein transport many years later, see: Isenman, L.D., C. Liebow, and S.S. Rothman. Protein transport across membranes—A paradigm in transition. *Biochim. Biophys. Acta*, 1241: 341–370, 1995.

(30) And then there was this curiosity. To successfully collect zymogen granules from homogenates, the tissue could *not* be suspended in an ionic medium of sodium and potassium salts at roughly neutral pH, the environment in which they happily reside in the cell. Homogenization performed in this "physiological" medium leads to the disappearance of the white granule sediment and the appearance of granule contents in the suspending medium. This was said to be due to lysis.

To successfully collect zymogen granules from tissue homogenates required an unnatural, unphysiologic suspending medium. It had to be both non-ionic and acidic. None of this made sense in terms of a lytic mechanism. Why would lysis occur in a neutral, polar medium and be prevented in an acidic, non-polar one? Why were granules stable in a non-physiological environment and unstable in a physiological one?

Moreover, release as a function of pH is differential. It is *not* the same for different enzymes housed in the same objects, pouring out of a fractured object in strictly proportional amounts like so many beads of different colors. See Rothman, S.S. The behavior of isolated zymogen granules: pH—dependent release and reassociation of protein. *Biochim. Biophys. Acta*, 214: 567–577, 1971.

## Chapter 16

(31) Singer, S.J. and G.L. Nicolson. The fluid mosaic model of the structure of cell membranes. *Science*, 175: 720–731, 1972.

(32) Wallach, D.F.H. and A.S. Gordon. Lipid protein interactions in cellular membranes. *Fed. Proc.*, 27: 1263–1268, 1968.

(33) Neurons might be equally unpromising in that their membranes must be highly resistive to carry an electric current, but that is another subject.

(34) With the exception of small, often functionally important but structurally more or less meaningless variations (called "allosteric"), protein structure derived from crystal diffraction was thought to be permanent, unchanging whatever the environment, except when all function is lost due to the massive disordering of the peptide chain called "denaturation" (for example, at acidic pH).

(35) There is a cleft on one side of enzyme molecules, a microscopic invagination that is called the active site. This is where they do their catalytic chemical work. But this site only accounts for a relatively small fraction of the total mass of the large molecule. What is the rest of the molecule up to, what does it do? The common view was that it was either functionless or the servant of the active site, shaping it to make it functional. Over the years we have learned that the rest of these enormous molecules can serve a variety of other functions. The capacity for transport across membranes being one of them.

(36) Genetic sequencing and methods such as mass spectrometry have made it clear that a named protein, say chymotrypsinogen, is not a single entity, but has many different incarnations (often called alleles). Perhaps some of them have something to do with proteins crossing membranes.

## Chapter 17

(37) A minor addendum is needed here. There is a third mechanism for the movement of molecules. It is the flow of fluid, the result of hydrodynamic pressure like the elevation that drives water and *its contents* down rivers and streams. Though over short distances, such as biological membranes, such processes are usually insignificant, osmotic forces produce water flow across them and the substances (solutes) dissolved in it (the solvent) can be pulled across the membrane by that flow. This is known as solvent drag. So, just as the flotsam and jetsam suspended in river water is dragged downstream by the water's flow, if the channels through which water travels across membranes are broad enough, proteins can be dragged along for the ride.

(38) Rothman, S.S. Enzyme secretion in the absence of zymogen granules. *Am. J. Physiol.*, 228: 1828–1834, 1975. Also, see #58.

## Chapter 18

(39) Save for the few granules being formed just as the radioactive substance is added to the medium.

(40) Rothman, S.S. and L.D. Isenman. The secretion of digestive enzyme derived from two parallel intracellular pools. *Am. J. Physiol.*, 226: 1082–1087, 1974.

(41) Kelly, R.B. Pathways of protein secretion in eukaryotes. *Science*, 230: 25–32, 1985.

(42) Smaller vesicles become unstable, break apart.

## Chapter 19

(43) Isenman, L.D. and S.S. Rothman. Diffusion-like processes account for digestive enzyme secretion by the pancreas. *Science*, 204: 1212–1215, 1979.

(44) Ho, J.J.L. and S.S. Rothman. The nature of the flow dependence of protein secretion by the exocrine pancreas. *Am. J. Physiol.*, 242: G32–G39, 1982.

(45) Melese, T. and S.S. Rothman. Distribution of three hexose derivatives across the pancreatic epithelium: paracellular shunts or cellular passage? *Biochim. Biophys. Acta*, 763: 212–219, 1983.

(46) Melese, T. and S.S. Rothman. Increased phosphate efflux from the acinar cell during protein secretion. *Am. J. Physiol.*, 245: C121–C124, 1983.

(47) Melese, T. and S.S. Rothman. The pancreatic epithelium is permeable to sucrose and inulin across secretory cells. *Proc. Nat. Acad. Sci. (USA)*, 80: 4870–4874, 1983.

## Chapter 20

(48) Ho, J.J.L. and S.S. Rothman. Origin of the decline in pancreatic amylase secretion with time in vitro. *Am. J. Physiol.*, 245: C21–C27, 1983.

# Chapter 21

(49) Liebow, C. and S.S. Rothman. Enteropancreatic circulation of digestive enzymes. *Science*, 189: 472–474, 1975.

(50) Goetze, H. and S.S. Rothman. Enteropancreatic circulation of digestive enzyme as a conservative mechanism. *Nature*, 257: 607–609, 1975.

(51) Isenman, L.D. and S.S. Rothman. Transport of alpha-amylase across the basolateral membrane of the pancreatic acinar cell. *Proc. Natl. Acad. Sci. (USA)*, 74: 4068–4072, 1977.

(52) Isenman, L.D. and S.S. Rothman. Transpancreatic transport of digestive enzyme. *Biochim. Biophys. Acta*, 585: 321–332, 1979.

(53) Miyakasa, K. and S.S. Rothman. Redistribution of amylase activity accompanying its secretion by the pancreas. *Proc. Nat. Acad. Sci. (USA)*, 73: 5438–5442, 1982.

In addition, for example,

(54) Goetze, H. and S.S. Rothman. Amylase transport across ileal epithelium in vitro. *Biochim. Biophys. Acta*, 512: 214–220, 1978, or

(55) Miyasaka, K. and S.S. Rothman. The endocrine secretion of alpha-amylase by the pancreas. *Am. J. Physiol.*, 241: G170–G175, 1981.

# Chapter 22

(56) Porter, K.R. and A.B. Novikoff. The 1974 prize for physiology or medicine. *Science*, 186: 516–520, 1974.

(57) Palade, G. Intracellular aspects of the process of protein synthesis. *Science*, 189: 347–358, 1975.

(58) Rothman, S.S. Protein transport by the pancreas. *Science*, 190: 747–763, 1975.

# Chapter 24

(59) Goncz, K.K., P. Batson, D. Ciarlo, B.W. Loo, Jr, and S.S. Rothman. An environmental sample holder for the X-ray microscope. *J. Microscopy (London)*, 168: 101–110, 1992.

(60) Goncz, K.K. and S.S. Rothman. Protein flux across the membrane of single secretion granules. *Biochim. Biophys. Acta*, 1109: 7–16, 1992.

(61) Kühne, W. and A.Sh. Lea. Beobachtungen über die Absonderung des Pankreas, *Untersuch. Physiol. Inst. Univ. Heidelberg*, 2: 448, 1882.

(62) Ermak, T.H. and S.S. Rothman. Zymogen granules decrease in size in response to feeding. *Cell Tiss. Res.*, 214: 51–66, 1981.

(63) Ermak, T. and S.S. Rothman. Increase in zymogen granule volume accounts for increase in volume density during prenatal development of pancreas. *Anat. Record*, 207: 487–507, 1983.

(64) Ermak, T.H. and S.S. Rothman. Large decrease in zymogen granule size in the postnatal rat pancreas. *J. Ultrastruct. Res.*, 70: 242–256, 1980.

# Chapter 25

(65) Adelson, J.W. and S.S. Rothman. Selective pancreatic enzyme secretion due to a new peptide called chymodenin. *Science*, 183: 1087–1089, 1974.

(66) Adelson, J.W. and S.S. Rothman. Chymodenin, a duodenal peptide: specific stimulaton of chymotrypsinogen secretion. *Am. J. Physiol.*, 229: 1680–1686, 1975.

(67) Tseng, H.C., J.H. Grendell, and S.S. Rothman. Regulation of digestion. II. Effects of glucagon and insulin on pancreatic secretion. *Am. J. Physiol.*, 246: G451–G456, 1984.

(68) Rothman, S.S. The molecular regulation of digestion: bond-specific and short-term. *Am. J. Physiol.*, 226: 77–83, 1974.

(69) Grendell, J.H., H.C. Tseng, and S.S. Rothman. Regulation of digestion. I. Effects of glucose and lysine on pancreatic secretion. *Am. J. Physiol.*, 246: G445–G450, 1984.

(70) Niederau, C., J.H. Grendell, and S.S. Rothman. Digestive end products release pancreatic enzymes from particulate cellular pools including zymogen granules. *Biochim. Biophys. Acta*, 881: 281–291, 1986.

(71) Rothman, S.S. The behavior of isolated zymogen granules: pH-dependent release and reassociation of protein. *Biochim. Biophys. Acta*, 214: 567–577, 1971.

(72) Rothman, S.S. The association of bovine $\alpha$-chymotrypsinogen and trypsinogen with rat zymogen granules. *Am. J. Physiol.*, 222: 1299–1302, 1972.

(73) Goncz, K.K. and S.S. Rothman. A trans-membrane pore can account for protein movement across zymogen granule membranes. *Biochim. Biophys. Acta*, 1238: 91–93, 1995.

## Chapter 26

(74) Rothman, S. The incoherence of the vesicle theory of protein secretion. *J. Theor. Biol.*, 245: 150–160, 2007.

(75) Rothman, S. The constancy of the internal environment: Proteins in plasma. *FASEB J.*, 19: 1383–1388, 2005.

(76) Rothman, S. Reply to the letter of Professors Brusch and Deutsch. *J. Theor. Biol.*, 252: 374–375, 2008.

(77) Other than as the result of an extremely unlikely chance occurrence.

(78) There is also a less fundamental, but no less challenging problem for secretion by the acinar cell of the exocrine pancreas by vesicle processes. The acinar cell secretes 10–20 different digestive enzymes. If we insist that this occurs by exocytosis, we have given ourselves another irresolvable difficulty. It turns out that the enzymes are stored together in the same granules in the same cells (see Reference 23), and yet are made at different rates. But their release by exocytosis can only occur at a single rate. Hence, even if we could magically balance synthesis and movement, balancing it for *each* of these proteins is just not possible.

## Chapter 27

(79) Johannes Poncius's commentary on John Duns Scotus's *Opus Oxoniense*, book III, dist. 34, q. 1. in John Duns Scotus *Opera Omnia*, vol. 15, Ed. Luke Wadding, Louvain (1639), reprinted Paris: Vives, (1894) p. 483a.

(80) Ariew, R. Ockham's Razor: A Historical and Philosophical Analysis of Ockham's Principle of Parsimony, 1976.

## Chapter 30

(81) The relatively small number of projects supported by private foundations usually drink from the same well of expert opinion.

Here are some other books and review articles on this subject written by my colleagues and myself. They should give you a sense of how our ideas evolved over the years.

(82) Rothman, S.S. Transport of protein by pancreatic acinar cells: Random or Selective. In *The Exocrine Glands*, eds. S.Y. Bothelo, F.P. Brooks, and W.B. Shelley (University of Pennsylvania Press, Philadelphia, PA, 1969, pp. 169–181).

(83) Rothman, S.S. The digestive enzymes of the pancreas, a mixture of inconstant proportions. *Ann. Rev. Physiol.*, 39: 373–389, 1977.

(84) Rothman, S.S. The passage of protein through membrane—old assumptions and new perspectives. *Am. J. Physiol.*, 238: G391–G402, 1980.

(85) Rothman, S.S. *Nonvesicular Transport* (John Wiley and Sons, New York, 1985).

(86) Rothman, S.S. The signal hypotheses. In: *Nonvesicular Transport*, eds. S.S. Rothman and J.J.L. Ho (Wiley, New York, 1985b, pp. 137–167).

(87) Rothman, S.S. *Protein Secretion: a Critical Analysis of the Vesicle Model* (John Wiley and Sons, New York, 1985).

(88) Rothman, S.S. The regulation of digestive reactions by the pancreas. In: *Handbook of Physiology, The Gastrointestinal System III*, ed. J.G. Forte (Oxford University Press, New York, 1989, pp. 465–476).

(89) Rothman. S.S., C. Liebow, and J. Grendell. Non-parallel transport and mechanisms of secretion. *Biochim. Biophys. Acta*, 1071: 159–173, 1991.

(90) Rothman, S.S. (Editor) *Membrane Protein Transport*. Volumes 1, 2 and 3 (JAI Press, Hampton Hill, Middx, U.K. and Greenwich, Conn., 1995–1996).

(91) Isenman, L.D., C. Liebow, and S.S. Rothman. Protein transport across membranes—A paradigm in transition. *Biochim. Biophys. Acta*, 1241: 341–370, 1995.

(92) Rothman, S.S. The sorting of proteins. In: *Principles of Medical Biology*, Volume 7A, Membranes and Cell Signaling, ed. E. Bittar (JAI Press/Elsevier, London, 1997, pp. 205–227).

(93) Rothman, S.S. *Lessons From the Living Cell—The Limits of Reductionism* (McGraw-Hill, New York, 2002).

(94) Rothman, S.S., C. Liebow, and L.D. Isenman. The conservation of mammalian digestive enzymes. *Physiol. Revs.*, 82: 1–18, 2002.

Here are a few additional research articles from our laboratory that are relevant to the subject:

(95) Rothman, S.S., J. Grendell, K. McQuaid, and J. Underwood. The relationship between the light scattering properties of zymogen granules and the release of their contained proteins, *Biochim. Biophys. Acta*, 1074: 85–94, 1991.

(96) Goncz, K.K. and S.S. Rothman. The protein content and morphogenesis of zymogen granules. *Cell Tiss. Res.*, 280: 519–530, 1995.

(97) Young, M.K., H.C. Tseng, H. Fang, W. Liang, and S.S. Rothman. Comparison of stored and secreted rat pancreatic digestive enzymes by mass spectrometry: I. α-amylase. *Biochim. Biophys. Acta*, 1293: 63–71, 1996.

(98) Rothman, S. How is the balance between protein synthesis and degradation achieved? *Theor. Biol. and Med. Mod.*, 7: 25, 2010.

# Index

Note: Page numbers followed by 'n' refer to end notes.